ISBN 978-0-484-27861-4
PIBN 10738345

For support please visit www.forgottenbooks.com

UNITED STATES NATIONAL MUSEUM BULLETIN 268

A Revision
of the
Genus Eucerceris Cresson
(Hymenoptera: Sphecidae)

HERMAN A. SCULLEN

SMITHSONIAN INSTITUTION PRESS
WASHINGTON, D.C.
1968

SMITHSONIAN

INSTITUTION

MUSEUM

OF

NATURAL

HISTORY

Errata

Page 65, line 2.   **For** FIGURES 41, 92 a, b, c, d, e, f, g, h **read** FIGURE 40
Page 66, line 19.   **For** FIGURE 40 **read** FIGURES 41, 92 a, b, c, d, e, f; g, h
Page 65, map legend.   **For** FIGURE 41.   Western U.S. *E. superba superba* Cresson
   read FIGURE 40.   Western U.S. *E. superba bicolor* Cresson
Page 67, map legend.   **For** FIGURE 40.   Western U.S. *E. superba bicolor* Cresson
   read FIGURE 41.   Western U.S. *E. superba superba* Cresson

Publications of the United States National Museum

The scientific publications of the United States National Museum include two series, *Proceedings of the United States National Museum* and *United States National Museum Bulletin.*

In these series are published original articles and monographs dealing with the collections and work of the Museum and setting forth newly acquired facts in the field of anthropology, biology, geology, history, and technology. Copies of each publication are distributed to libraries and scientific organizations and to specialists and others interested in the various subjects.

The *Proceedings*, begun in 1878, are intended for the publication, in separate form, of shorter papers. These are gathered in volumes, octavo in size, with the publication date of each paper recorded in the table of contents of the volume.

In the *Bulletin* series, the first of which was issued in 1875, appear longer, separate publications consisting of monographs (occasionally in several parts) and volumes in which are collected works on related subjects. *Bulletins* are either octavo or quarto in size, depending on the needs of the presentation. Since 1902, papers relating to the botanical collections of the Museum have been published in the *Bulletin* series under the heading *Contributions from the United States National Herbarium.*

This work forms number 268 of the *Bulletin* series.

FRANK A. TAYLOR
Director, United States National Museum

U.S. GOVERNMENT PRINTING OFFICE
WASHINGTON : 1968

For sale by the Superintendent of Documents, U.S. Government Printing Office
Washington, D.C. 20402 - Price 50 cents (paper cover)

Contents

A Revision
of the
Genus Eucerceris Cresson
(Hymenoptera: Sphecidae)

Introduction

The wasp genus *Eucerceris* was erected by E. T. Cresson in 1865. In 1939 the writer published his first extensive paper on this genus (Scullen, 1939). A second paper with a revised key to the known species (Scullen, 1948) was published later describing additional species. Since the above publications appeared, the writer, through the assistance of grants from the National Science Foundation and the Graduate School of Oregon State University, has been able to study types in this country and in the museums of Europe. Furthermore, in more recent years a vast amount of additional material and information has been secured through the field work of the writer and numerous other workers. Many new species have been discovered in Mexico. The distribution data and information relative to known species have been greatly extended by recent workers.

Structural Characters

The genus *Eucerceris* which is limited to North America may be distinguished from the more widely distributed genus *Cerceris* by the following structural characters:

(1) The 3rd submarginal cell of the fore wing is more or less inflated in *Eucerceris* and the second submarginal cell is not always petiolate as in *Cerceris*.

(2) The hair lobes of the male clypeus are less distinctly separated from the other setae of the lower face and are never "waxed" as is so true of *Cerceris*.

(3) The terga of *Eucerceris* show a more or less distinct mesal depression in which the punctation is crowded. This depressed area is inclined to be darker than the convex surfaces of the terga.

(4) The male pygidium of *Eucerceris* differs markedly from that of the female and shows distinct lateral denticles.

The genotype of *Eucerceris* was designated as *Eucerceris fulvipes* Cresson by V. S. L. Pate (1937, p. 27).

Biological Studies

Very little is yet known about the biology of most species of *Eucerceris*. In the writer's 1939 publication *Eucerceris flavocincta* Cresson was recorded as collecting *Dyslobus lecontei* Casey at Breitenbush Hot Springs in Marion County, Oreg. Bohart and Powell (1956) reported *E. flavocincta* Cr. as carrying in an undescribed species of *Dyslobus*. Linsley and MacSwain (1954) report *E. ruficeps* Scullen using the weevils *Dysticheus rotundicollis* Van Dyke and *Sitona californicus* Fahrens in the sand dune region east of Antioch on the lower San Jaquine River. Krombein (1960a, b) reported *E. triciliata* (now known to be *E. pimarum* Rohwer, Scullen, 1965) as taking the weevil, *Minyomerus languidus* Horn near Portal, Ariz. *E. rubripes* Cresson was found by Mont A. Cazier collecting a species of weevil, *Peritaxia* sp., 2 mi. NE. of Portal, Ariz., July 28, 1961.

Acknowledgments

Material studied in the preparation of this publication has come from most of the insect collections found on this continent and several European institutions as mentioned in the author's publication on *Cerceris* north of the Mexican border (Scullen, 1965b). Also, many individual collectors have contributed to these studies as mentioned in the above-cited publication. The drawings in this publication were prepared by Thelwyn M. Koontz. Special acknowledgment should go to Janet Bedea who has done most of the typing and many other routine services in this connection.

The writer is also deeply indebted to Dr. Paul O. Ritcher, Head of the Entomology Department, Oregon State University, for the use of facilities in that department. Special acknowledgment should go to Dr. Karl V. Krombein of the United States National Museum who has checked over the keys and assisted in many other ways to make this and related publications possible.

Financial assistance for these studies has come from grants by the National Science Foundation and from grants for General Research under the Graduate School, Oregon State University.

Key to Species of *Eucerceris* Cresson

Seven segments in abdomen; 13 segments in antennae . . . . male
Six segments in abdomen; 12 segments in antennae . . . . female

MALES

1. Second submarginal cell not petiolate 2
   Second submarginal cell petiolate 32
2. No distinct row or cluster of erect bristles on venter (western United States and Canada) . . . . **flavocincta** Cresson
   One or more distinct rows or clusters of bristles on venter 3
3. Row or cluster of bristles on 5th sternum only, sometimes very inconspicuous 4
   Rows or clusters of bristles on more than one sternum 13
4. Mid femora with a deep depression beneath at base bordered by dense long setae; two inconspicuous clusters of bristles on sternum 5 (one or both clusters sometimes broken off) 5
   Mid femora normal in form 6
5. Terga of abdomen with broad fulvous bands showing little or no emargination (southern Arizona and New Mexico, western Texas, north central Mexico) . . . . **lacunosa lacunosa** Scullen
   Tergal bands deeply emarginate with ferruginous (Coachuila, Mexico, southern Texas, and one record from Portal, Ariz.)
   lacunosa sabinasae, new subspecies
6. Scutum covered with a dense layer of short setae giving a distinct velvet-like appearance (western Mexico, Sonora to Oaxaca) . . **velutina** Scullen
   Setae of thorax normal 7
7. Black with yellow to white markings 8
   Partly or largely ferruginous and fulvous 11
8. Light vittae of face fusing above the antennae (southwestern states)
   arenaria Scullen
   Light vittae of face not fusing above the antennae 9
9. Dark vittae of face do not extend beyond the dorsal clypeal border, except for a hair line between the clypeal lobes; single row of bristles on 5th sternum (Rocky Mountain area) . . . . **fulvipes** Cresson
   Dark vittae of face extending onto clypeus 10

10. Row of bristles on 5th sternum subequal in length to one third the width of the medial clypeal lobe and divided medially; 3rd tergum immaculate (southern Mexico). . . . . **melanosa** Scullen
    Row of bristles on 5th sternum subequal in length to width of medial lobe of clypeus; 3rd tergum with a narrow band (southwestern states and northern Mexico) . . . . **melanovittata** Scullen

11. Row of bristles on 5th sternum subequal in length to one third the width of sternum, not divided medially (Rocky Mountain area S. to Chihuahua, Mexico) . . . . **rubripes** Cresson
    Length of bristle row on 5th sternum not over one fourth the width of the sternum; bristle row divided medially. 12

12. Medial lobe on apex of pygidium extends well beyond lateral denticles (western Texas, southern New Mexico, and Chihuahua, Mexico) . . **mellea** Scullen
    Medial lobe on apex of pygidium does not extend beyond apices of the lateral denticles (south central Texas south into Coahuila and Nueva Leon, Mexico) . . . . **sinuata** Scullen

13. Three subequal rows or clusters of bristles on the venter 14
    Rows or clusters of bristles on 5th sternum, if present, are either much shorter or different in form from those on other sterna. 16

14. Each row of bristles distinctly divided into two parts (central plains and eastern states) . . . . **zonata** (Say)
    Bristles in undivided rows. 15

15. Length about 15 mm.; scutum immaculate (Rocky Mountain area, southern Canada to Arizona and New Mexico) . . . . . **superba superba** Cresson
    Length about 10 mm.; usually two short yellow stripes on the scutum (southwestern states, southern California, east to western Texas, north central Mexico) . . . . **pimarum** Rohwer

16. Bristles of the 5th sternum somewhat shorter than the others, the apical row itself shorter than the other rows, and the bristles of the apical row "waxed" into a compact layer 17
    Bristles of the 5th sternum very short, forming very short rows, evanescent or absent, and, if present, not "waxed" 22

17. Black with yellow to white markings, no amber or brown 18
    Not all black with yellow or white markings, some amber or brown markings 19

18. Markings yellow; enclosure strongly ridged and with two oval patches; bands on terga 2, 3, 4, and 5 separated into basal and apical parts by the black depressed areas (central California and western Nevada)
    . . . . insignis Provancher
    Markings creamy white; enclosure very weakly ridged and immaculate; complete bands on apical ridges of terga 2, 3, 4, and 5 (recorded only from state of San Luis Potosí, Mexico) . . . . **atrata** new species

19. Background color largely black but with ferruginous bordering the yellow markings and on the genae (central Mexico). . **zimapanensis**, new species
    Background color largely ferruginous to fuscous 20

20. Dorsum of thorax ferruginous to fulvous with yellow markings (central plains from Montana S. into central Mexico, W. to southern California and E. to central Texas) . . . . **canaliculata canaliculata** (Say)
    Dorsum of thorax fuscous to dark fuscous with yellow markings 21

21. (Scattered records throughout the southwestern states and into central Mexico) . . . . canaliculata atronitida Scullen [1]
(Recorded only from 32 miles SE. of Guaymas, Mexico)
sonorae, new species [1]

22. Complete broad bands on terga 1 to 5, covering basal as well as apical ridges 23
Basal ridges and depressed areas of three or more terga largely black, but apices of terga banded 27

23. Scutum immaculate (Bay area of central California) . . ruficeps Scullen
Yellow or creamy white wedge-shaped patches or stripes on the anterior portion of the scutum 24

24. Markings creamy white 25
Markings yellow 26

25. Enclosure immaculate (western Nevada) . . . . . elegans elegans Cresson
Enclosure with two small oval patches (Mojave Desert area of California
mojavensis, new species

26. Background color of thorax and abdomen black (east central area of California) . . . . elegans monoensis, new subspecies
Background color of thorax and abdomen largely ferruginous (southwestern desert areas, southern California to western Texas and north to Nebraska)
apicata Banks

27. Short row or cluster of short bristles on 5th sternum 28
No distinct row or cluster of bristles on 5th sternum 31

28. Enclosure relatively smooth with oval creamy white or yellow patches; light vittae of face may or may not fuse above the antennae; body surface glossy
29
Enclosure immaculate and distinctly ridged over entire surface 30

29. Light vittae of face usually convergent above antennae; small evanescent lateral spots on the scutum; markings light yellow to cream (southeastern Oregon, southern Idaho, southwestern Wyoming) . . . barri, new species
Light vittae of face broadly separated above antennae; scutum immaculate; markings yellow (southern Oregon, northern California) . . similis Cresson

30. Ferruginous area back of compound eye; creamy white vittae of face do not fuse above the antennae (Baja California, Mexico) . . . . baja Scullen
Black and creamy white only, no ferruginous back of the eye; light vittae of face fuse above the antennae (Baja California, Mexico) . . pacifica Scullen

31. Legs dark ferruginous to fuscous; enclosure immaculate; body black with creamy white markings, no ferruginous (central plateau area of Mexico, one record each from western Texas and southern New Mexico)
morula morula, new species
Legs ferruginous; terga 1 and 2, and all of sterna more or less infused with ferruginous; evanescent small spots of cream on enclosure (south central New Mexico and western Texas) . . . morula albarenae, new subspecies

32. Third tergum largely ferruginous (southwestern states, north central Mexico) . . . . tricolor Cockerell
All terga black with yellow to cream-colored markings 33

---

[1] The males of *canaliculata atronitida* Scullen and *sonorae* Scullen at present are indistinguishable except when associated with the females.

33. Scape broad and flattened; denticles on the posterior apical angles of the first
    five segments of the flagellum (southern Arizona, southwestern New
    Mexico, Sonora and Baja California, Mexico) . . . . . **angulata** Rohwer
    Antennae normal in form 34
34. Enclosure ridged; dark vittae of face extending to the free border of the
    clypeus 35
    Enclosure pitted; lower face without dark vittae 36
35. Fore and mid femora with oval creamy white patches (north central
    Mexico) . . . . **baccharidis,** new species
    Femora usually immaculate (arid sections W. of Rocky Mountains)
                                                    vittatifrons Cresson
36. Thorax and first two abdominal segments immaculate; mandibles normal
    (southern Mexico, states of Puebla and Oaxaca) . . **stangei,** new species
    Thorax and first two abdominal segments with yellow markings; mandibles
    unusually broad (Rocky Mountains, Montana south to central Mexico)
                                                    montana Cresson

FEMALES

1. Second submarginal cell not petiolate 2
   Second submarginal cell petiolate 8
2. Without a distinct projection or extension of the ventral clypeal margin or
   surface, except for a slight medial carina; ferruginous with yellow
   markings 3
   With a conspicuous projection or extension of the ventral clypeal margin or
   surface 4
3. Abdominal terga with broad yellow bands showing little or no emargination
   (southern Arizona, southern New Mexico, western Texas and north central
   Mexico) . . . . **lacunosa lacunosa** Scullen
   Abdominal terga with distinctly emarginate bands of yellow (southern
   Arizona, southern Texas, Coahuila, Mexico)
                                            lacunosa sabinasae, new subspecies
4. Clypeal surface with an emarginate projection on the surface of which are two
   subparallel carinae; brown with sclerite borders and some depressed areas
   black (known only from the states of Hidalgo, Michoacan and Jalisco,
   Mexico) . . . . **brunnea** Scullen
   Ventral clypeal border with a rounded medial extension 5
5. Mandibular denticles acute; with a large amount of fulvous to ferruginous
   coloring (western to southwestern Mexico from Sonora to Oaxaca)
                                                    velutina Scullen
   Mandibular denticles rounded; black with yellow markings only 6
6. Length about 23 mm. (only one specimen from Panama, Central America)
                                                    violaceipennis Scullen
   Length about 15 mm 7
7. Two lateral large yellow spots on the scutellum; enclosure black except for
   two elongate yellow spots; tergal bands much reduced (one specimen only
   from Temax, north Yucatan, Mexico)   **punctifrons punctifrons** (Cameron)
   Emarginate band on the scutellum; entire enclosure yellow; solid but narrow
   bands on all terga (one specimen from El Salvador)
                                            punctifrons cavagnaroi, new subspecies
8. Distinct projections or elevations on the surfaces of the lateral clypeal
   lobes 9
   All projections, elevations, or processes confined to the medial clypeal
   lobe and/or margins of the lateral lobes 11

9. A distinct truncate cone-shaped elevation on the frons just above the epistomal suture . . . . **sonorae,** new species
No elevation on the frons above the epistomal suture 10

10. Ferruginous with yellow markings (mostly in the Southwest but recorded from central California to the Missouri River and from Montana to central Mexico) . . . . **canaliculata canaliculata** (Say)
Fuscous to black with yellow markings (scattered records in Utah, southern Arizona, western Texas, and central Mexico)
canaliculata atronitida Scullen

11. Enclosure densely and completely punctate, not ridged 12
Enclosure smooth or ridged; if punctures are present, they do not cover entire enclosure 14

12. With a pair of blunt denticles on the margin of the medial lobe of the clypeus; 1st and 2nd terga immaculate (known only from the states of Puebla and Oaxaca in Mexico) . . . . **stangei,** new species
With a single broad rounded process on the medial clypeal lobe; 1st and 2nd terga with light markings 13

13. Narrow creamy white bands on the terga (southern Arizona, southern New Mexico, Baja California, and Sonora, Mexico) . . . . **angulata** Rohwer
Broad yellow bands on terga (Rocky Mountains, Mont. S. to central Mexico)
montana Cresson

14. Distinct but often small elevations on the surface of the medial clypeal lobe (not just a convex surface) 15
No distinct elevation or denticle on the surface of the medial clypeal lobe 22

15. Mandibles with a single subbasal or medial tooth on inner margin . . . 16
Mandibles with two subbasal or median teeth on inner margin 19

16. Five denticles on the medial clypeal border, the mesal one smaller than the others; considerable amber on most of the dorsal parts (Chisos Mountains, Tex., limited records from New Mexico and Chihuahua, Mexico)
mellea Scullen
Less than five denticles on the medial clypeal border 17

17. Four distinct subequal denticles on the medial clypeal margin and these subequally spaced (Rocky Mountain states from Montana to the Mexican border) . . . . **fulvipes** Cresson
Two widely separated denticles on the clypeal border between which there is usually a bicuspid medial extension 18

18. Considerable amber on dorsal parts . . . . **rubripes** Cresson
Amber markings, if any, largely on ventral parts; medial extension not always bicuspid . . . . **melanovittata** Scullen

19. The more basal denticle of the mandible much the larger of the two; largely amber and yellow (widely scattered over the Southwest and into northern Chihuahua, Mexico) . . . . **apicata** Banks
The more basal denticle of the mandible much the smaller; two widely separated denticles on the clypeal margin; body black and creamy white . . 20

20. Pygidium noticeably narrowing basally (arid sections west of the Rocky Mountains) . . . . **vittatifrons** Cresson
Pygidium not noticeably narrowing basally 21

21. A small blunt single elevation on the surface of the medial lobe of the clypeus (Southwestern states, California to New Mexico) . . . **arenaria** Scullen

Elevation on the surface of the medial lobe of the clypeus broad and tending to be bidentate (recorded only from the state of San Luis Potosi, Mexico) **atrata** Scullen

22. Apical clypeal margin with one large process 23
    Apical clypeal margin with more than one process or denticle and these relatively small 26

23. Process on the clypeal margin not acute (one specimen only recorded from the type locality which is given as "Mexico") . . **cerceriformis** Cameron
    Process on the clypeal margin acute 24

24. Mandibles with one large obtuse—angulate denticle somewhat divided (Central and northeastern states from Colorado to New England states) **zonata** (Say)
    Mandibles with one single acute denticle 25

25. Abdomen largely yellow (Rocky Mountain area, Alberta to western Texas) **superba superba** Cresson
    Basal terga of abdomen ferruginous; apical terga more or less fuscous to black (scattered records in the northern Rocky Mountains area from Alberta south to western Kansas) . . . . **superba bicolor** Cresson

26. Two completely separated single, acute, mandibular denticles, one above the other 27
    All mandibular denticles on the dorsal carina 28

27. Fulvous to ferruginous, without black (south central Texas and northeastern Mexico) . . . . sinuata Scullen
    Black with yellow markings, very limited fulvous replacing the yellow (known only from the state of Oaxaca, Mexico) . . . **menkei,** new species

28. Two closely placed truncate denticles on the clypeal margin with a cluster of bristles between them (southwestern states and north central Mexico) **pimarum** Rohwer
    Two widely separated acute, or more than two denticles on the clypeal margin 29

29. Lateral clypeal denticles, approximately at the junction of the medial and lateral clypeal lobes, undivided and unidentate 30
    Lateral clypeal denticles either bidentate or divided into two appressed denticles 35

30. Dorsal body parts black with creamy white or yellow markings, no brown or amber markings 31
    Dorsal body parts with brown, red, or amber 34

31. Markings yellow; enclosure smooth; mandibular denticles undivided (west of the Rocky Mountains in the United States) . . . similis Cresson
    Markings creamy white; enclosure ridged; mandibular denticle bicuspidate 32

32. With a small, single, mesal denticle just above the median teeth on margin of medial clypeal lobe (state of San Luis Potosi, Mexico and adjoining areas) . . . . baccharidis, new species
    With a pair of medial denticles a short distance above apical margin of the medial clypeal lobe 33

33. Venter fuscous to black (mostly known from central plateau of Mexico, rarely from Texas and New Mexico) . . . morula morula, new species
    Venter ferruginous to fulvous (south central New Mexico and western Texas) **morula albarenae,** new subspecies

34. Approximate length 10 mm.; basal segments of abdomen only ferruginous (Arizona, New Mexico, western Texas, and central plateau of Mexico) .... **tricolor** Cockerell
   Length 15 mm.; entire body and legs fulvous with yellow markings (Baja California, Mexico) .... **lapezensis,** new species
35. Black with yellow or creamy white markings, no brown or amber 36
   With considerable brown or amber replacing the black background coloration at least on the head 38
36. Length approximately 15 mm.; enclosure finely ridged, immaculate (Rocky Mountains and western states) .... **flavocincta** Cresson
   Length 10 mm. or less; enclosure with two light oval patches 37
37. Markings yellow; double yellow bands on two or more terga; enclosure coarsely ridged (California, western Nevada, northern Baja California, Mexico) .... **insignis** Provancher
   Markings creamy white; single bands on all terga; enclosure smooth or nearly so; lateral clypeal denticles may be bidentate or tridentate becoming sinuate when badly worn (southwestern Oregon, southern Idaho, southern Wyoming) .... **barri,** new species
38. Almost entire body ferruginous to fulvous; light markings very limited or absent (desert areas of southern California, north central and southern Nevada, and northern Baja California, Mexico) . . **ferruginosa** Scullen
   Amber or brown markings more limited black with prominent yellow or creamy white markings 39
39. Markings creamy white; considerably ferruginous on the thorax and basal two terga (western Nevada) ...... **elegans elegans** Cresson
   Thorax and abdomen black and yellow only 40
40. Yellow markings dull, not glossy; without yellow spots on enclosure (lower Sacramento area in California) .... **ruficeps** Scullen
   Yellow markings very glossy; often with oval yellow spots on enclosure (Mono and Inyo counties, Calif.) . . . **elegans monoensis,** new subspecies

1. *Eucerceris angulata* Rohwer

FIGURES 1, 59 a, b, c, d, e, f, g, h, i, j

Eucerceris angulata Rohwer, 1912, p. 326.—Scullen, 1939, pp. 17, 18, 56–58, figs. 34, 35, 36, 48, 51, 75, 89, 108, 122, 134b, 156; 1948, pp. 157, 158, 180; 1951, p. 1011.

The female of *E. angulata* Rohwer was described by Rohwer in 1912. The male was described and the female redescribed by Scullen (1939).

In later years both males and females have been taken by several collectors in numerous locations as listed below.

TYPE.—The holotype female of *Eucerceris angulata* Rohwer is at the American Museum of Natural History. The type female is from Lower California, between San Jose del Cabo and Triunfo. Collected by 'Albatross.'

DISTRIBUTION.—Southern Arizona, southern New Mexico in the United States; Baja California and Sonora, Mexico. Specimens are as follows:

ARIZONA: ♂, 5 mi. SW Apache, Cochise County, Aug. 17, 1961, *Baccharis glutinosa* (M. A. Cazier); ♂, Arivaipa, Aug. 25, 1953 (Bryant); ♂, Benson, 3,700 ft., July 26, 1946 (H. A. Scullen); ♂, Continental, Aug. 12, 1957 (G. D. Butler); ♂, Douglas, July 19, 1950 (R. H. Beamer); ♂, 6 mi. NE. Douglas, Aug. 11, 1940, *Eriogonum* (C. D. Michener); ♂, 10 mi. NE. Douglas, Aug. 11, 1940 (E. S. Ross); ♂, 16 mi. E. Douglas, Cochise County, Sept. 8, 1958 (P. D. Hurd); ♂, Madera Canyon, Santa Rita Mts., July 31, 1958 (R. M. Bohart); 2♂♂, Mt. Lemmon Road, Santa Catalina Mts., 3,500 ft., Aug. 15, 1954 (R. M. Bohart); ♂, 6 mi. N., Pearce, Aug. 6, 1955, *Lepidium* (G. Butler, Z. Noon), ♀, 25 mi. E., Pearce, July 29, 1954 (Butler-Werner); 2♂♂, 30 mi. E., Pearce, July 10, 1955, *Sapindus saponaria* (G. Butler, F. Werner); ♂, Portal, Chiricahua

Figure 1. Western U.S. *E. angulata* Rohwer

Mts., June 29, 1956 (O. L. Cartright); 2♂♂, same locality, Aug. 3, 9, 1958 (R. M. Bohart); ♂, same locality, July 8, 1963 (A. Raske); ♂, 1.5 mi. NE., Portal, Cochise County, 5,000 ft., Aug. 8, 1959 (E. G. Linsley); ♀, 24♂♂, 2 mi. NE., Portal, Cochise County, July 29, 30, Aug. 1, 3, 5, 1959, *Baccharis glutinosa* (M. Statham); ♀, ♂, same locality, Aug. 2, 1959 (E. G. Linsley); ♀, 5♂♂, same locality, July 2, 1960 (M. A. Cazier); 18♂♂, same locality, Aug. 13, 14, 15, 17, 18, 1962, *Baccharis glutinosa* (H. A. Scullen); 3♀♀, 10♂♂, same location, July 14, 18, Aug. 24, 1964, July 27, 1965, *Baccharis glutinosa* (J. Puckle, M. A. Cazier, Mortenson); 4♂♂, 2.5 mi. NE., Portal, Cochise County, Aug. 5, 9, 13, 1959 (M. A. Cazier); 3♀♀, 2♂♂, Sabino Canyon, Santa Catalina Mts., Pima County, Aug. 11, 13, 14, 1924 (E. P. Van Duzee); ♂, same locality, 1952 (L. D. Beamer); 3♂♂, same locality, July 9, 1952 (R. H. & L. D. Beamer); ♀, Southwest Research Station, Portal, 5,400 ft., June 29, 1956 (H. A. Scullen); ♂, 12 mi. S., Stafford, Graham County, 4,250 ft., Sept. 14, 1962 (H. A. Scullen); 2♂♂, Tucson (F. H. Snow); 2♂♂, 16 mi. S., Tucson, Aug. 11, 1924 (J. O. Martin); ♀, 18 mi. S. Tucson, July 13, 1924 (E. C. Van Dyke); ♀, 8 mi. N. Tucson, Pima County,

June 11, 1964 (J. M. Davidson). NEW MEXICO: ♀, 5♂♂, 1 mi. N., Rodeo, July 28, 29, 1963 (Cazier and Mortenson); ♀, ♂, 18 mi. N., Rodeo, Hidalgo County, Aug. 25, 1958 (R. M. Bohart). MEXICO: BAJA CALIFORNIA: ♂, Big Canyon, Sierra Laguna, Oct. 13, 1941 (Ross and Bohart); 2♂♂, Cabo San Lucas, July 17, 1959 (H. B. Leech); 2♂♂, Coyote Cove, Concepción Bay, Oct. 1, 1941, (Ross and Bohart); ♂ La Paz, June 29, 1919 (G.F.Ferris); ♀, same locality, Oct. 1923 (W.M. Mann); ♂, same locality, Oct. 10–12, 1954 (F. X. Williams); 3♀♀, 6♂♂, 10 mi. E. of San Ignacio, Sept. 30, 1941, Composite (Ross and Bohart); ♀, San José del Cobo to Triunfo, 1911; 9♂♂, San Pedro, Oct. 7, 1941, Compositae (Ross and Bohart). SONORA: 7♀♀, 16♂♂, 32 mi. SE., Guaymas, 125 ft., Sept. 14, 1963, *Baccharis glutinosa* (Scullen and Bolinger); 2♂♂, 10 mi. E., Navojoa, Aug. 13, 1959 (W. L. Nutting, F. G. Werner).

PREY RECORDS: None.

PLANT RECORDS: *Acacia* sp. (Arizona), *Baccharis glutinosa* (Arizona, Sonora, Mex.), Compositae (Baja California, Mex.), *Eriogonum* sp. (Arizona), *Haplopappus* (*Aplopappus*) (Arizona), *Koeberlinia spinosa* (New Mexico), *Lepidium* sp. (Arizona), *Sapindus saponaria* (Arizona).

2. Eucerceris apicata Banks

FIGURES 2, 60a,b,c,d,e,f,g

Eucerceris apicata Banks, 1915, p. 404. ♂.—Scullen, 1965, pp. 132–135.
Eucerceris elegans Mickel (!) 1916, p. 413; 1917, pp. 454, 456.—Scullen, 1939, pp. 18, 19, 32–34, figs. 21, 41, 63, 81, 96, 114, 128, 143; 1948, pp. 156, 159, 171; 1951, p. 1011.
Eucerceris conata Scullen, 1939, pp. 18, 34–35, figs. 22, 23, 64, 97, 144, ♀; 1948, pp. 158, 171, 172; 1951, p. 1011.
Eucerceris hespera Scullen, 1948, pp. 156, 171–172, figs. 8A, B, C, 14, ♂; 1951, p. 1012.

TYPES.—The holotype male of *E. apicata* Banks is at the Museum of Comparative Zoology, Harvard, No. 13792. It is from Yuma, Ariz. The holotype female of *E. conata* Scullen is at the University of Nebraska. It was taken at Halsey, Nebr., Aug. 28, 1911 (J. T. Zimmer). The holotype male of *E. hespera* Scullen is at the California Academy of Sciences and was taken 25 mi. E. of El Paso, Tex. on U.S. Hwy. 62, July 13, 1942 (H. A. Scullen). A discussion of the confused relationship of this and closely allied species was published by the writer in 1965.[1]

DISTRIBUTION.—Abundant in southern Arizona, New Mexico, and western Texas with scattered records as far north as southwestern South Dakota and west to eastern California. Specimens are as follows:

ARIZONA: ♂, 32 mi. SE. Ajo, Pima County, Sept. 2, 1959 (G. I. Stage); ♂, 11 mi. E. Moenkopi, Coconino County, 5,200 ft., July 27, 1937 (Rehn, Pate, Rehn); ♂, Quijotoa, Pima County, Aug. 27, 1927; 2♂♂, Toreva, Navajo County, Aug. 30, 1911 (S. O. Barrett); ♂, Tucson, Oct. 27, 1937, at *Cuscuta umbellata*

---

[1] See note under *E. rubripes* Cresson, p. 55.

(R. H. Crandall); ♂, Tucson, July 17, 1955 (G. D. Butler); ♂, Willcox, Cochise Country, July 31, 1951 (C. W. O'Brier). CALIFORNIA: 3♂♂, Benton, Mono County, July 25, 1942 (W. M. Pearce). COLORADO: ♂, St. Charles River, July 27, 1948 (C. H. & D. Martin). NEBRASKA: 2♀ ♀, 2♂♂, Halsey, Thomas County, July 25, 1912, Aug. 28, 1911, Aug. 29, 1912 (J. T. Zimerman); ♂, same locality, Aug. 14, 1920 (C. B. Phillip); ♂, same locality, Aug. 12, 1925 (L. C. Worley); ♀, same locality, Aug. 13, 1925 (R. W. Dawson); ♂, Bridgeport, July 11, 1927 (C. E. Mickel). NEW MEXICO: ♂. Albuquerque, July 18, 1902 (Oslar); 2♀ ♀, Las Cruces, 3,880 ft., June 18, 1942 (H. A. Scullen); ♂, same locality, June 19, 1942 (E. C. Van Dyke); ♂, 17 mi. W. Las Cruces, Dona Ana County, 4,400 ft., June 26, 1956 (H. A. Scullen); ♂, "New Mexico." SOUTH DAKOTA: ♂, Hot Springs, July 10, 1924. TEXAS: ♀, Clint, El Paso County, Oct. 7, 1943 (R. W.

Figure 2. Western U.S. *E. apicata* Banks

Strandtmann); Dalhart, Dallam County, June 8, 1950 (R. H. Beamer); ♀, El Paso, July 11, 1942 (E. C. Van Dyke); ♀, same locality, June 23, 1947 (Spieth) ♀, 2♂♂, same locality, Sept. 27, 1943 (R. W. Strandtmann); 73♀ ♀, 42♂♂, 10–20 mi. E. of El Paso, June 21, 22, 1942 (H. A. Scullen); 3♀ ♀, 15 mi. N. El Paso, June 23, 1942 (H. A. Scullen); 26♀ ♀, 30♂♂, same locality, June 22, 1942 (E. C. Van Dyke); 10♀ ♀, 6♂♂, 10 mi. E. El Paso, June 21, 1942 (E. C. Van Dyke); 2♂♂, 20 mi. N. El Paso, June 19, 1942 (H. A. Scullen); 4♀ ♀, 3♂♂, same locality, June 19, 21, 1942 (E. C. Van Dyke); 6♂♂, 25 mi. E. El Paso, July 13, 1942 (H. A. Scullen); 2♂♂, same locality, July 13, 1942 (E. C. Van Dyke); ♀, ♂, Van Horn, June 24, 1942 (E. C. Van Dyke). WYOMING: ♂, Torrington, July 30, 1939 (J. Standiah). MEXICO: CHIHUAHUA: ♂, Samalayuca, June 24, 1947 (G. M. Brandt).

PREY RECORDS: None.

PLANT RECORDS: *Cuscuta umbellata* (Arizona). *Wislizenia* (Arizona).

3. *Eucerceris arenaria* Scullen

FIGURES 3, 61 a, b, c, d, e, f, g

Eucerceris arenaria Scullen, 1948, pp. 156, 157, 168–170, figs. 6a, b, c, d, e, f, 15; 1951, p. 1011.

TYPES.—The holotype female from Helendale, Calif. and the allotype male from Cuchinbury Springs, Calif., Aug. 16, 1937, on *Solidago confinus* (P. H. Timberlake) are at the University of California at Riverside, Calif.

DISTRIBUTION.—In recent years the known distribution of *E. arenaria* Scullen has been greatly extended. Its known range extends

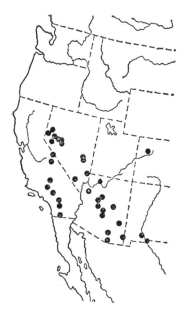

Figure 3. Western U.S. *E. arenaria* Scullen

from southern California through Nevada and Arizona to southwestern New Mexico, with one record from Colorado. Specimens are as follows:

ARIZONA: ♀, Bartlett Dam, July 13, 1960 (J. E. Gillaspy); ♂, Chiricahua Mts., Cochise County, Aug. 11, 1961 (W. J. Hanson); ♂, Dewey, Yavapai County, Aug. 2, 1960 (M. L. Rice); ♀, North Rim, Grand Canyon, 8,000–9,000 ft. elev., July 18, 1938 (F. E. Lutz); ♂, Granite Reef Dam, Maricopa County, May 8, 1964 (J. H. Puckle); ♂, 23 mi. E. Kingman, July 1, 1952 (Beamer and party); ♂, Nogales, July 12, 1913 (Oslar); ♂, Phoenix, Oct. 20, 1934 (R. H. Crandall); 7♂♂, Portal, Cochise County, Aug. 3, 9, 12, 1958 (R. M. Bohart); 19♀♀, 5♂♂, same locality, Aug. 5, 9, 12, 15, 1958 (P. D. Hurd); 2♀♀, ♂, 2 mi. NE. Portal, Cochise County, Aug. 2, 1959 (E. G. Linsley); 10♀♀, 46♂♂, same locality, July 29, 30, Aug. 1, 3, 5, 1960 (M. Statham); ♂, same locality, Sept. 19, 1961, at *Euphorbia albomarginata* (M. Cazier); 3♀♀, 37♂♂, same locality, Aug. 13, 14, 15, 17, 18, 1962, at *Baccharis glutinosa* (H. A. Scullen); 2♀♀, same locality, 4,600 ft. elev., Aug. 15, 1962 (J. Wilcox); ♂, Sabino Canyon, Pima

County, May 5, 1955 (G. D. Butler); ♂, Salt River Mts., 1,300 ft. elev., May 9, 1926 (A. A. Nichol); ♂, 12 mi. S. Sedona, Oak Creek Canyon, July 18, 1957 (C. W. O'Brien); ♀ Hwy. 93, 4 mi. SE. Sta. Maria River, Yavapai County, July 21, 1960 (J. E. Gillaspy); ♂, 12 mi. S. Stafford, Graham County, 4,250 ft. elev., Sept. 14, 1962 (H. A. Scullen); ♂, Tucson (F. H. Snow); ♀, 10 mi. E. Wickenburg, May 30, 1958 (J. C. Hall. CALIFORNIA: ♂, 25 mi. E. Barsto, San Diego County, June 30, 1952 (Beamer and party); 5 ♂ ♂, Borego, San Diego County, May 2, 1952, May 5, 1956, at *Croton californicus* (P. D. Hurd); 2 ♂ ♂, same locality, May 2, 1952 (J. G. Rozen); 2 ♀ ♀, same locality, May 27, 28, 1955 (R. O. Schuster); ♀, 3 ♂ ♂, Borego State Park, San Diego County, Apr. 27, 1950 (C. D. MacNeill); 2 ♂ ♂, same locality, July 2, 1953 (M. Wasbauer); ♀, same locality, Apr. 24, 1954 (J. G. Rozen); ♀, 2 ♂ ♂, same locality, Apr. 28, 1955, at *Chaenactis* sp. (P. D. Hurd); ♂, Cajon, San Bernardino County, July 17, 1956 (H. R. Moffitt); ♂, Deep Creek, San Bernardino County, Aug. 11, 1956 (J. Ciffall); 2 ♂ ♂, same locality and date (E. I. Schlinger); ♂, Deep Springs, Inyo County, Sept. 10, 1956 (J. A. Chemsak); 5 ♂ ♂, same locality, Oct. 10, 1956 (E. G. Linsley); ♂, Fish Creek, Imperial County, Aug. 10, 1955 (J. E. H. Martin); ♀, 5 ♂ ♂, 25 mi. S. Ivanpah, San Bernardino County, Oct. 13, 1958 (J. W. MacSwain); 4 ♂ ♂, Jacumba, San Diego County, Aug. 12, 1935 (E. I. Beamer); ♂, Johannesburg, Kern County, Sept. 4, 1958 (E. I. Schlinger); ♀, 5 ♂ ♂, Kramer, San Bernardino County, Sept. 4, 1958 (E. I. Schlinger); ♀ Lone Pine Creek, Inyo County, June 6, 1939 (R. M. Bohart); ♀, 3 ♂ ♂, Magnesia Canyon, Riverside County, June 25, July 2, 1952 (A. T. MacClay); 2 ♀ ♀, 2 ♂ ♂, same locality, June 28, 1952 (E. G. Linsley); 6 ♂ ♂, same locality, July 2, 1952 (S. Miyagawa); ♀, same locality, June 22, 1958 (E. I. Schlinger); 2 ♂ ♂, Mazowrka Canyon, Inyo County, July 2, 1953 (J. W. MacSwain); ♀, Mint Canyon, July 8, 1956 (Simonds); ♀, Mojave June 8 1938 (F. T. Scott); ♂ 6 mi. E. Mojave Kern County Oct. 6 1957 (J. W. MacSwain); ♀, 4 ♂ ♂, 7 mi. E. Mojave, Kern County, Sept. 30, 1963 (R. Wescott); 12 ♀ ♀, 17 ♂ ♂ Morongo Valley, San Bernardino County, July 19, 26, Aug. 10, Oct. 5, 12, 1958, at *Eriogonum inflatum* (O. C. La France); ♂, Palm Springs, Riverside County, June 24, 1952 (Beamer and party); ♂, Parker Dam, San Bernardino County, Apr. 26, 1949 (C. D. MacNeill); ♂, Scissors Crossing, San Diego County, Oct. 4, 1955 (J. T. Powell); ♀, Tahquitz Canyon, near Palm Springs, Riverside County, Apr. 22, 1963 (Stange and Parker); 10 ♀ ♀, 23 ♂ ♂, same locality, June 8, 1957 (Menke, Stange, and Bromley); ♀, Thousand Palms, Apr. 17, 1955 (W. R. Richards); ♂, Walker Pass, Kern County, Sept. 21, 1957 (H. R. Moffitt). COLORADO: ♂, New Castle, Garfield County, Aug. 5, 1956. NEVADA: 2 ♀, 10 ♂ ♂, Alamo, Lincoln County, June 25–26, July 12, 31, 1958 (F. D. Parker); ♀, ♂, 6 mi. N. Alamo, Lincoln County, July 2, 1958 (F. D. Parker); 2 ♀ ♀, ♂ 11 mi. W. Eastgate, Churchill County, Aug. 11, 1958 (E. G. Linsley); 3 ♂ ♂, 23 mi. E. Fallon, Churchill County, June 20, 1958 (J. W. MacSwain); 3 ♂ ♂, 25 mi. NE. Fernley, Aug. 29, 1956 (T. R. Haig); ♂, 4 mi. S. Gabbs, Nye County, July 11, 1961; ♂, 3 mi. W. Hazen, Churchill County, June 20, 1961 (F. D. Parker); ♂, 8 mi, E. Hiko, Lincoln County, July 20, 1958 (F. D. Parker); ♂, Nixon, Washoe County, July 3, 1962 (M. E. Irwin); ♂, Nixon, Washoe County, July 3, 1962 (R. M. Bohart); Nixon, Washoe County, July 3, 1962 (R. J. Gill); ♂, 1 Nixon, Washoe County, June 21, 1960 (F. D. Parker); ♂, 2 mi. N. Nixon, Washoe County, June 23, 1961 (F. D. Parker); 6 ♂ ♂, 8 mi. S. Overton, Clark County, June 5, 1960 (R. W. Lauderdale); ♂, Pahrump, Nye County, July 25, 1958 (R. C. Bechtel); ♂, 15 mi. E. Reno, July 15, 1962 (F. D. Parker); ♂, 15 mi. E. Reno, July 15, 1962 (M. E. Irwin); 2 ♂ ♂, Riverside, Clark County, July 21, 1952 (M. Cazier, W. Gertsch, R. Schrammel); 6 ♀ ♀, 24 ♂ ♂, Riverside, Clark County, Aug. 21, 1958, *Eriogonum, Croton californicus* (R. W. Lauderdale); ♀, 5 mi. NE.

Smith, Lyon County, Aug. 23, 1960 (J. A. Chemsay); ♂, Walker Lake, July 29, 1957; Willow Creek Camp, Charleston Mts., July 1, 1954 (J. W. MacSwain). NEW MEXICO: 2 ♀ ♀, S. Cienega Peak, Peloncillo Mts., 4,500 ft., Aug. 27, 1937 (Rehn, Pate, Rehn); 2 ♀ ♀, 2 ♂ ♂, Granite Pass, Hidalgo Co., Aug. 22, 25, 1958 *Croton* sp. (P. D. Hurd); ♀, Rodeo, Hidalgo Co., Aug. 7, 1958 (G. B. Pitman); ♀, same locality and date (C. G. Moore), ♂, same locality, Aug. 23, 1958 (M. A. Cazier); ♀, same locality, July 28, 1963 (Cazier and Mortenson); 9 ♂ ♂, same locality, July 29, 1963, *Koeberlinia spinosa* (Cazier and Mortenson); 4 ♂ ♂, 18 mi. N. Rodeo, Aug. 25, 1958 (G. B. Pitman, R. E. Rice, P. M. Marsh); 4 ♂ ♂, same locality and date (R. M. Bohart); TEXAS: ♂, El Paso, June 23, 1947 (Spieth).

PREY RECORDS: None.

PLANT RECORDS: *Baccharis glutinosa* (Arizona). *Chrysothamnus* sp. (Nevada). *Cleome lutea* (Nevada). *Croton californicus* (California). *Dalea polyadenia* (Nevada). *Eriogonum* sp. (Arizona, California, New Mexico, Nevada). *Eriogonum inflatum* (California). *Euphorbia albomarginata* (Arizona). *Koeberlinia spirosa* (New Mexico). *Lepidospartum squamatum* (California). *Prosopis juliflora* (Arizona). *Solidago confinis* (California). *Tetradymia canescens* (Nevada).

4. Eucerceris atrata, new species

FIGURES 4, 62 a, b, c, d, e, f

FEMALE.—Length 13 mm. Body black with creamy white markings, legs largely ferruginous; punctation average; pubescence short.

Head slightly wider than the thorax; black except for large eye patches, most of the clypeus, the area between the antennal scrobes, and a large patch back of the eye, all of which are creamy white; clypeal border medially depressed with an angular denticle between the medial and lateral clypeal lobes, above and more mesad there is a smaller denticle, and a cluster of bristles on the meson; a small convex elevation on the medial clypeal lobe; mandibles with one prominent medial denticle and a smaller one more basad; antennae normal in form.

Thorax black except for a band on the pronotum, a band on the scutellum, the metanotum, a large patch on the propodeum, a small patch on the pleuron and the posterior lobe of the pronotum, all of which are creamy white; tegulae low and smooth; enclosure smooth except for a medial groove and radiating short ridges medially, otherwise smooth; mesosternal tubercles small but distinct; legs ferruginous except for small spots of white on the fore and mid femora; wings subhyaline except for a lightly clouded anterior margin; second submarginal cell petiolate.

Abdomen black except a divided band on tergum 1, emarginate bands on terga 2, 3, 4, and 5, and lateral small patches on sterna 3 and 4, all of which are creamy white; pygidium with sides converging apically to a moderately acute apex.

MALE.—Length 13 mm. Body black with creamy white markings, legs largely ferruginous; punctation average; pubescence average.

Head subequal in width to thorax, black except for a large lateral eye patch, the clypeus and an elongate patch between the antennal scrobes, all of which are creamy white; clypeal border with three subequal black denticles; mandibles without denticles, ferruginous with darker apices; antennae black.

Thorax black except the pronotum, a small patch on the posterior lobe of the pronotum, band on the scutellum, the metanotum, large patches on the propodeum, a triangular patch on the pleuron and a small patch on the tegulae, all of which are creamy white; tegulae low and smooth; enclosure smooth except for a medial groove and

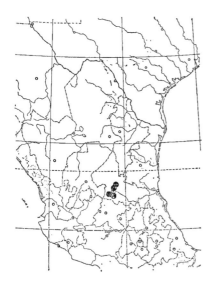

Figure 4. Central Mexico. *E. atrata*, new species

evanescent grooves basad; mesosternal tubercles absent; legs ferruginous except for a small creamy white spot on each of the fore and mid femori and the basal parts of all legs which are dark fuscous to black; wings subhyaline with a clouded area on the anterior part apically.

Abdomen black except for a broad band constricted medially on tergum 1, a broad band deeply emarginate on tergum 2, narrow bands widening laterally on terga 3, 4, and 5, emarginate band on tergum 6, broken band on sternum 2, band on sternum 3 and lateral patches on sternum 4, all of which are creamy white; prominent rows of bristles on sterna 3 and 4, a shorter row of short appressed bristles on sternum 5; pygidium as illustrated (fig. 62f).

TYPES.—The ♀ type and ♂ allotype were taken 17 mi. NE. of San Luis Potosí, S.L.P., Mexico, Sept. 6, 1963, at 6,200 ft. elev.

on *Baccharis glutinosa* (H. A. Scullen and Duis Bolinger). The type female and allotype male are at the U.S. National Museum, no. 69225.

PARATYPES.—MEXICO: State of San Luis Potosí: 7 ♀ ♀, 21 ♂ ♂, 17 mi. NE. of San Luis Potosi, Sept. 6, 1963, 6,200 ft. elev., at *Baccharis glutinosa* (H. A. Scullen and D. Bolinger); 5 ♂ ♂, same locality, Oct. 3, 1957 (H. A. Scullen); 7 ♂ ♂, 10 mi. SW., San Luis Potosí, 7,300 ft. elev., Oct. 1, 2, 1957 at *Baccharis glutinosa* (H. A. Scullen); ♂, 31 mi. NE., San Luis Potosi, 5,500 ft. elev., Oct. 3, 1957 (H. A. Scullen); ♂, 19 mi. SW. San Luis Potosi, 7,200 ft. elev., Oct. 4, 1963 at *Baccharis glutinosa* (H. A. Scullen and D. Bolinger); 4 ♂ ♂, 4 mi. SW., San Luis Potosi, 6,500 ft. elev., Sept. 4, 1963, at *Baccharis glutinosa* (H. A. Scullen and D. Bolinger).

DISTRIBUTION.—Known only from the state of San Luis Potosí, Mexico.

PREY RECORD.—None.

PLANT RECORD.—*Baccharis glutinosa*.

5. *Eucerceris baccharidis*, new species

FIGURES 5, 63 a, b, c, d, e, f

FEMALE.—Length 9 to 10 mm. Black with creamy white markings and some leg parts fuliginous; punctation average; pubescence short.

Head slightly wider than the thorax; black except frontal eye patches, a patch on the medial and each lateral lobe of the clypeus, an elongate patch between the antennal scrobes, an elongate area back of the eye and a spot on the base of the mandible, all of which are creamy white; clypeal border with a broadly based denticle at the junction of the medial lobe and each lateral lobe, more acute denticles at the meson, a 5th smaller denticle above the medial pair, a cluster of five or six bristles above the medial pair and below the medial single denticle; surface of the medial lobe of the clypeus slightly convex; mandibles with a prominent blunt denticle with a smaller and more acute and basad denticle fused with it at its base; antennae black and normal in form.

Thorax black except for a solid band on the pronotum, two patches on the scutellum, a solid band on the metanotum, an oval patch on the propodeum, a patch on the pleuron below the wing, four small spots on the sternum mesad of the second and third coxae and a small spot on the tegulae, all of which are creamy white; tegulae low and smooth; enclosure with a medial groove and ridged laterally with the ridges subparallel to the matenotum but bending posterolaterally; mesosternal tubercles appearing as indistinct elevations; legs black to the apical ends of the first two femori and the third coxae, with prominent oval creamy white patches near the apical ends of the first two pair of femori, the remaining more apical segments of all legs are fuliginous; wings subhyaline with the anterior margins dark, second submarginal cell petiolate.

Abdomen black except for a divided band on tergum 1, narrow solid bands on the apical half of terga 2, 3, 4, and 5, all of which are creamy white; venter immaculate; pygidium with a broad base, sides slightly convex and converging to a slender but rounded apex.

MALE.—Length 9 to 10 mm. Black with creamy white markings except for some leg parts which are fuliginous; punctation average; pubescence short.

Head subequal in width to the thorax; black except for elongate frontal eye patches which are medially emarginate, most of the medial clypeal lobe, large triangular patches on the lateral lobes, an elongate area between the antennal scrobes, a round spot back of the eye and a patch on the base of the mandible, all of which are

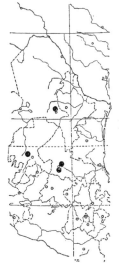

Figure 5. Central Mexico. *E. baccharidis*, new species

creamy white; clypeal border with three black denticles on the medial lobe, the medial one the largest; hair lobes indistinct; mandibles without denticles; antennae black and normal in form.

Thorax colored as for the female; tegulae low and smooth; enclosure as in the female; mesosternal tubercles absent; legs colored as in the female; wing colors and venation as for the female.

Abdomen colored dorsally as on the female; venter with solid creamy white bands on sterna 3 and 4, sternum 5 with minute spots laterally; pygidium as illustrated (fig. 63f).

Both sexes of this species closely resemble *E. vittatifrons* Cresson in size, colors, color pattern and wing venation. The females, however, differ in the structure of the clypeal borders and the oval white patches on the first and second pair of femora of *E. baccharidis*. The males may usually be separated by the oval patches on the fore and mid femora of the latter species.

TYPES.—The type ♀ and allotype ♂ were taken 17 mi. NE. of San Luis Potosí, S. L. P., Mexico, at about 6,200 ft. elev. on *Baccharis glutinosa*, Sept. 6, 1963 (H. A. Scullen and Duis Bolinger). The type female and allotype male of *E. baccharidis* Scullen are at the U.S. National Museum, no. 69226.

PARATYPES.—MEXICO: ♀, 1.5 mi. S. Fresnillo, Zac., Aug. 6, 1954 (E. G. Linsley, J. W. MacSwain, R. F. Smith); ♂, 4 mi. SW. San Luis Potosi, 6,500 ft. elev., Sept. 4, 1963 (Scullen and Bolinger); 62 ♀ ♀, 91 ♂ ♂ 17 mi. NE. of San Luis Potosí, S. L. P., Sept. 6, 1963, 6,200 ft. elev., on *Baccharis glutinosa* (Scullen and Bolinger); ♂, 15 mi. N. Saltillo, Coah., 4,450 ft. elev., Sept. 9, 1963 (Scullen and Bolinger).

DISTRIBUTION.—State of San Luis Potosí and one record from Coahuila, all from Mexico.

PREY RECORD.—None.

PLANT RECORD.—*Baccharis glutinosa.*

6. *Eucerceris baja,* Scullen

FIGURES 6, 64 a, b, c

Eucerceris baja Scullen, 1948, pp. 170–171.

MALE.—The male only was described in 1948. The following specimens have been examined by the author since the above publication:

MEXICO: Baja California: 2 ♂ ♂, 10 mi. E. of Bahía San Quintin, Sept. 9, 10, 1955 (F. X. Williams); 12 ♂ ♂, 28 mi. SE. of Arco, Rancho Santa Marguerita, July 3, 4, 1960 (A. E. Michelbacher).

The female is still unknown.

TYPE.—The holotype male from 20 mi. N. of Mesquital, Baja California, Mexico, Sept. 27, 1941 (Ross and Bohart), is at the California Academy of Sciences.

DISTRIBUTION.—Known only from the records published by the author in 1948 and above from Baja California.

PREY RECORDS.—None.

PLANT RECORDS.—None.

7. *Eucerceris barri,* new species

FIGURES 7, 65 a, b, c, d, e, f, g

FEMALE.—Length 10 mm. Black with creamy white markings; punctuation somewhat less crowded than average; pubescence sparse and very short.

Head slightly wider than the thorax, black except for irregular frontal eye patches, a medial spot above the antennal scrobes and a slightly elongate spot back of the eye, all of which are creamy white; clypeal border with low single denticles on the medial lobe near the fusion with the lateral lobes, a low carina connects these two lateral denticles, below this ridge there is a medial cluster of bristles, below these bristles there is a broad extension of the clypeal border the lateral angles of which appear as distinct denticles, surface of the clypeal medial lobe slightly convex; mandibles with a single bicuspidate (cusps appear as one in some specimens) denticle which is low and with a broad base; antennae immaculate, normal in form.

Thorax black except for a divided band on the pronotum, a band on the scutellum, a band on the metanotum, oval spots on the enclosure, patches on the propodeum, small spots on the tegulae and small spots on the pleuron, all of which are creamy white; tegulae low and smooth; enclosure with a deep medial groove, lightly ridged at about 45° with its base and with a pair of oval spots variable in size; mesosternal tubercle absent; legs black except for traces of

Figure 6. Baja California. *E. baja* Scullen

creamy white on the tarsal segments; wings subhyaline with the anterior part clouded, second submarginal cell not petiolate.

Abdomen black except for a divided band on tergum 1, emarginate bands on terga 2, 3, 4, and 5; venter immaculate; pygidium with sides very slightly convex and converging to a rounded apical tip.

MALE.—Length 9 mm. Black with creamy white markings; punctation somewhat less crowded than average; pubescence sparse and very short.

Head slightly wider than the thorax, black except for the face and an elongate patch on the genae bordering the eye, both of which are creamy white; narrow black stripes extend from the antennal scrobes to connect by an evanescent line with the black of the vertex; base of mandibles, patch on the scape and basal segments of the flagellum creamy white; clypeal border with three blunt denticles on the medial lobe, the medial denticle the smallest; mandibles without denticles but with a slightly elevated carina; antennae normal in form.

Thorax black except for a solid band on the prothorax, the posterior lobe of the pronotum, two spots laterally on the scutum, a band on the scutellum, the metanotum, two oval spots on the enclosure, large patches on the propodeum, a large area on the pleuron, most of the sternum, the tegulae, all of which are creamy white; tegulae low and smooth, enclosure smooth except for a medial groove and light rugae in the lateral angles; mesosternal tubercles absent; legs largely creamy white except for elongate black areas on the dorsal surfaces of all segments except the tarsi; all tarsi become more or less ferruginous apically; wings subhyaline but lightly clouded along the anterior area, second submarginal cell not petiolate (fig. 65e).

Figure 7. Western U.S. *E. barri*, new species

Abdomen black except for broad bands on terga 1, 2, 3, 4, 5, and 6, those on 3, 4, and 5 confined medially to the apical part of the tergum, bands on sterna 2, 3, and 4, and evanescent lateral spots on sternum 5, all of which are creamy white; prominent rows of bristles on the apical margins of sterna 3 and 4, an almost indistinguishable pair of bristle clusters on sternum 5; pygidium average in form.

In some females the band on the metanotum is broken into two lateral patches. Some specimens of each sex have the markings somewhat yellow. In some females the 2nd submarginal cell is fully petiolate. On the males the light markings above the antennae may or may not be fused. This species is very close to *E. similis* Cresson

TYPES.—The ♀ type was taken at Jacob's Cabin, Heart Mt., Lake County, Oreg., 6,600 ft. elev., July 16, 1937 (Bolinger and Jewett). The ♂ allotype was taken at Sublett, Cassia County, Idaho, July 24, 1957 at *Achillea* (W. F. Barr). The type female and allotype male of *E. barri* Scullen are at the U.S. National Museum no. 69227.

PARATYPES.—CALIFORNIA: ♀, 4 mi. S. Ravendale, Lassen County, Aug. 10, 1959 (J. A. Chemsak). COLORADO: ♂, Maybell, Aug. 30, 1931 (R. H. Beamer). IDAHO ♂, Mackey, 5,899 ft. elev., July 28, 1957, at *Chrysothamnus* sp. (W. F. Barr); ♂, Montpelier, 6,100 ft. elev., July 6, 1920 (F. 4739 about 42°19′′N, 111°18′′W); 3♂♂, St. Anthony, Fremont County, July 17, 1956 (W. F. Barr); 2♀♀, ♂, Sublett, Cassia County, July 24, 1957 at *Achillea* (W. F. Barr); ♂, 12 mi. NE. Terreton, Jefferson County, July 17, 1956 (W. F. Barr). WYOMING: 2♂♂, Albany County, Aug. 11, 1949 (R. R. Driesbach and R. K. Schwab); ♀, same locality, Aug. 12, 1952 (R. R. Driesbach); ♀, 2♂♂, 25 mi. S. Bitter Creek, Sweetwater County, July; ♂, Green River (D. Elden Beck); ♂, Jackson, 6,300 ft. elev., July 13–17, 1920 (F. 4746 about 43°30′′N, 110°46′′W); ♂, Summit, 8,835 ft., Aug. 16, 1940 (H. E. Milliron); Summit, Albany County, 8,500 ft. elev., Aug. 10, 1950 (R. R. Driesbach); 4♂♂, Uinta County., 7,000 ft. elev., July 25, 1953 (R. R. Driesbach); ♂, Yellowstone Nat. Park- Aug. 7, 1930. UTAH: ♂, Randolph, July 25, 1953 (R. R. Driesbach).

DISTRIBUTION.—Lake County, Oreg., southern and eastern Idaho, southern Wyoming, northern Utah, northwestern Colorado, and northeastern California.

PREY RECORDS.—None.

PLANT RECORDS.—*Achillea* (Yarrow) (Idaho). *Chrysothamnus* sp. (Idaho).

8. *Eucerceris brunnea* Scullen

FIGURES 8, 66 a, b, c, d, e, f

Eucerceris brunnea Scullen, 1948, pp. 157, 159–160, figs. 1A, 1B, 1C, 13; 1957, pp. 155–156.

TYPE.—The holotype female of *E. brunnea* Scullen from Jacala, Hidalgo, Mexico, 4,500 ft. elev., June 22, 1936, (Ralph Haag), is at the Museum of Comparative Zoology, Harvard, no. 31258.

DISTRIBUTION.—This large and rare species was described in 1948 from a single female specimen taken at Jacala, Hgo., Mexico, 4,500 ft. elev., June 22, 1936 (Ralph Haag). Since that date three additional female specimens have been determined by the writer as follows: Guadalajara, Jal., 5,000 ft. elev., July 14, 1959 (H. E. Evans); 7 mi. S. of Cerro Teguila, Jal., 6,600 ft. elev., July 8, 1950 (Andrew Browne); 6 mi. NW. of Quiroga, Mich., July 11, 1963 (F. D. Parker, L. A. Stange). All are from Mexico.

PREY RECORDS.—None.

PLANT RECORDS.—None.

9a. *Eucerceris canaliculata canaliculata* (Say)

FIGURES 9, 10, 67 a, b, c, d, e, f, g

Philanthus canaliculatus Say, 1823, p. 80. ♂.—LeConte, 1883 Vol. 1 pp. 111, 167.
Cerceris bidentata Say, 1823, p. 80. ♀.—LeConte, 1883, vol. 1, p. 168.—Cresson, 1865, p. 130.
Eucerceris canaliculatus Cresson, 1865, p. 112.—Packard, 1866, p. 59.—Patton, 1879, pp. 357–359; 1880, p. 398.—Snow, 1881, p. 99.—Cresson, 1882, pp. vi, vii, viii; 1887, p. 281.—Ashmead, 1890, p. 32; 1899, p. 295.—Bridwell, 1899, p. 209.—Viereck and Cockerell, 1904, pp. 84, 85, 88.—Smith, H. S., 1908, p. 371.—Mickel, 1917, pp. 454, 455.—Scullen, 1939, pp. 18, 47–50, figs. 30, 45, 71, 85, 104, 118, 132, 151; 1948, pp. 156, 157, 179–180; 1951, p. 1011; 1964, p. 208.
Cerceris canaliculatus Schletterer, 1887, p. 488.—Dalla Torre, K. W. von, 1890, p. 200.
Cerceris cameroni Schulz, 1906, p. 194 (New name)

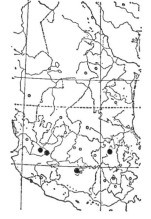

Figure 8. Central Mexico. *E. brunnea* Scullen

TYPE.—The Neotype male from Kansas is in the collection of the Philadelphia Academy of Natural Sciences. The homotype of *Cerceris bidentata* Say has been lost.

The male of *E. canaliculata* (Say) is at present indistinguishable from the male of *E. sonorae* Scullen except by their distribution or association with the females.

The subspecies *atronitida* Scullen is a darker form found in some areas. An uncommon very light form possibly should be considered as a subspecies. It is found in the lower desert areas such as Imperial County, Calif.

DISTRIBUTION.—The known distribution of the nominate subspecies has been greatly expanded since the writer's 1939 paper (p. 47). Several collecting trips into Mexico by the writer and others in recent years have shown it to range over much of northern Mexico and as far south as San Luis Potosí and San Blas, Nay. It is recorded as far north as Montana and west to the California coast. A record from Berkeley, Calif. may be open to question. In altitude it ranges from 1,500 to 6,500 ft.

PREY RECORDS.—It seems strange that as common as this species is there is no known case where it has been found nesting and the beetles used by it for prey are still unknown.

PLANT RECORDS.—*Acacia angustissima* (Arizona). *A. greggii* (catclaw) (Arizona). *Aplopappus* (=*Haplopappus*) *hartwigi* (Arizona). *Asclepias* sp. (Arizona, New Mexico, Texas). *Baccharis glutinosa* (Arizona, Mexico, New Mexico, Texas). *Baileya multiradiata* (Arizona). *B. pauciradiata* (New Mexico). *Cevallia sinuata* (Mexico).

Figure 9. Western U.S. *E. canaliculata canaliculata* (Say)

Chaenactis sp. (California). *Chrysopsis hirseetissima* (New Mexico). *Condalia lycioides* (Arizona, Texas). *Croton corymbulosus* (Texas). *Eriogonum annum* (Texas). *Gaillardia pulchella* (New Mexico). *Gutierrezia lucida* (New Mexico). *G. micrecephala* (Arizona). *Haplopappus* (see *Aplopappus*). *Hymenothrix wislizeni* (New Mexico). *Koeberlinia spinosa* (New Mexico). *Lepidium* sp. (Arizona, New Mexico). *Melilotus alba* (New Mexico). *Parthanium incanum* (Arizona). *Polygonum* sp. (New Mexico). *Robinia* sp. (Texas). *Senecio* sp. (New Mexico). *Salsola pestifer* (New Mexico). *Sapindus saponaria* (Arizona). *Tamarix gallica* (Arizona, California, New Mexico, Oklahoma, Texas). *Thelesperma megapotamicum* (Arizona, Colorado, New Mexico). *Wislizenia refracta* (Arizona, California, Texas). *Ximensia* (*Verbesina*) *encelioides* (Texas).

9b. *E. canaliculata atronitida* Scullen

FIGURES 11, 12

E. canaliculata atronitida Scullen, 1939, p. 50, fig. 152; 1951, p. 1011.
E. biconica Scullen, 1948, pp. 178–179; 1951, p. 1011.

E. canaliculata atronitida Scullen was described as a new variety by the writer in 1939. It is now being considered a subspecies. This darker form appears usually at higher altitudes in different areas. There may be some question about the advisability of considering it a subspecies.

TYPES.—The type male of *E. canaliculata atronitida* from Beaver Canyon, Utah, is in the U.S. National Museum, no. 51383. The

Figure 10. Western Mexico. *E. canaliculata canaliculata* (Say)

female type of *E. biconica* Scullen from 15 mi. N. of El Paso, Tex., June 23, 1942 (H. A. Scullen) is at the California Academy of Sciences.

DISTRIBUTION.—High elevations from Utah and western Colorado S. to central Mexico. Specimens are as follows:

ARIZONA: 2♂♂, Cornville, May 28, 1954 (G. D. Butler); ♀, 6 mi. E. Douglas, Aug. 11, 1940 at *Eriogonum* sp. (C. D. Michener); 2♂♂, Dragoon, Cochise County, July 20, 1917 (J. Bequaert). COLORADO: ♂, Frutia, 4,500 ft. elev., July 16, 1919; ♀, Palisade, 4,700 ft. elev., August 1932 (Lee Jeppson). NEW MEXICO: ♂, Albuquerque, 5,000 ft. elev., June 27, 1931 (Don Prentiss); ♀, 15 mi. N. Roswell, Chaves County, Aug. 13, 1964 (J. G. & R. L. Rozen). TEXAS: ♂, El Paso, June 23, 1942 (E. C. Van Dyke); ♀, 15 mi. N. El Paso, El Paso County, June 23, 1942 (H. A. Scullen); ♂, Shafter Lake, And. County, June 17, 1961 (J. E. Gillaspy); ♀, Sierra Blanca, El Paso County, July 8, 1917 (J. Bequaert); ♂, same locality, July 9, 1917 (R. C. Shannon); 5♂♂, Terlingua, Brewster County, May 10, 1927 (J. O. Martin); ♀, Valentine, Presidio County, July 8, 1917 (J. Bequaert). UTAH: 2♀♀, 4♂♂, Beaver Canyon; ♂, Leamington, July 16, 1949 (G. F. Knowlton); ♂, Lucin, Box Elder County, July 30, 1963 (P. Torchio, E. Bohart); ♂, Parogonah, July 24, 1951 (G. F. Knowlton and G. E. Bohart); 2♂♂, Parowan Canyon, Iron County, July 18, 1919; ♂, Trout Creek, Juab County, July 14, 1922 (Tom Spalding). MEXICO: Durango: 6♂♂, 40 and 45 mi. NW. Gomez Palacio, 3,700 and 3,800 ft. elev., Sept. 10, 1963 (Scullen and Bolinger). Hidalgo: 8♂♂, Zimapan, June 11–14, 1951, at *Eysenhardtia polystachya* (Ort.), (P. D. Hurd). Queretaro: ♀, ♂, 4 mi. N. Queretaro, 6,500 ft. elev., Sept. 19, 1963 (Scullen and Bolinger). San luis potosí: ♀, ♂, El Huizache, 4,500 ft. elev., Aug. 22, 1954, at *Lorre tridentata glutinosa;* ♀, 10 mi. NE. San Luis Potosi, 6,200 ft. elev., Aug. 22, 1954 (R. R. Driesbach); ♂, 18 mi. W. of S. L. P., 6,000 ft. elev., Oct. 3, 1957 (H. A. Scullen); ♀, ♂, 14 mi. E. of S. L. P., 6,200 ft. elev., Oct. 3, 1957 (H. A. Scullen); 4♀♀, 3♂♂, 15 mi. E. of S. L. P., 6,500 ft. elev., Oct. 3, 1957 (H. A. Scullen); 2♂♂, 17 mi. W. of S. L. P., July 24, 1962; ♂, 5.2 mi. N. of S. L. P. (Huizache Jc.), 4,200 ft. elev., Sept. 6, 1962; 3♀♀, 7♂♂, 40 mi. S. of S. L. P., 5,700 ft. elev., Sept. 5, 1963 (Scullen and Bolinger); 9♂♂, 5 mi. N.

Figure 11. Southwestern U.S. *E. canaliculata atronitida* Scullen

Figure 12. Central Mexico. *E. canaliculata atronitida* Scullen

of S. L. P., July 6, 1965 (C. H. Martin); 26♀♀, 23♂♂, 17 mi. N. of S. L. P., 6,200 ft. elev., Sept. 6, 1963 (Scullen and Bolinger); ♀, 24 mi. S. of S. L. P., July 7, 1965 (C. H. Martin).

Prey record.—None.
Plant record.—*Eysenhardtia polystachya* (Mexico). *Lorre tridentata glutinosa* (Mexico).

10. *Eucerceris cerceriformis* Cameron

Eucerceris cerceriformis Cameron, 1890, p. 130. ♀.—Ashmead, 1899, p. 295.—Scullen, 1939, pp. 18, 46; 1948, p. 158.
Cerceris cerceriformis Dalla Torre, C. G., 1897, p. 455.

Type.—The type of *E. cerceriformis* Cameron has not been located.
Distribution.—This species is known only from the type locality as given by Cameron.—"Hab. Mexico (coll. Saussure)."
Prey record.—None.
Plant record.—None.

11a. *Eucerceris elegans elegans* Cresson

FIGURES 13, 68 a, b; c

Eucerceris elegans Cresson, 1879, p. xxiii, ♂.; 1882, pp. vi, vii; 1887, p. 281.—
Ashmead, 1899, p. 295.—Cresson, 1916, p. 100.—Mickel, 1916, p. 413;
1917, pp. 454, 456.—Scullen, 1939, pp. 32–34; 1948, pp. 156, 159, 171; 1951,
p. 1011; 1965, pp. 132–135.
Cerceris elegantissima Schletterer, 1887, p. 490.—Dalla Torre, C. G., 1897, p. 458.
Cerceris nevadensis Dalla Torre, K. W. von, 200–201.

A revised synonymy of *E. elegans* Cresson and closely related species (*E. apicata* Banks and *E. pimarum* Rohwer) has recently been published by the writer (Scullen, 1965, pp. 132–135).

TYPE.—The holotype male of *E. elegans* Cresson is at the Philadelphia Academy of Natural Sciences, no. 1968. It was taken in Nevada.

Figure 13. Western U.S. *E. elegans elegans* Cresson

DISTRIBUTION.—Western Nevada. (Record as given in my 1965 paper, pp. 134–5.)

NEVADA: ♀, Dayton, Lyon Co., June 28, 1959 (T. Haig); 2 ♂ ♂, 7 mi. N. Dyer, Esmeralda County, July 2, 1958, at *Melilotus alba* (R. C. Bechtel); 2 ♂ ♂, 23 mi. E. Fallon, Churchill County, June 20, 1958, at *Dalea polyadenia* (J. W. MacSwain); ♂, same locality, June 20, 1958 (E. G. Linsley); ♀, Fernley, Lyon County, Aug. 15, 1953 (R. M. Bohart); ♂, 4 mi. NE. Fernley, Lyon County, May 30, 1958 (T. R. Haig); ♂, Nixon, Washoe County, June 20, 1927 (E. C. Van Dyke); 4 ♂ ♂, same locality, June 21, 1960, June 24, 1961 at *Tetradymia canescens* (F. D. Parker); ♀, 5 ♂ ♂, same locality, June 22, July 3, 1962 (R. J. Gill); 2 ♂ ♂, same locality, June 22, 1962 (R. M. Bohart); ♂, same locality, July 3, 1962 (M. E. Irwin); ♂, same locality, Aug. 14, 1963 (E. J. Montgomery); ♀, Pyramid Lake, Washoe County, July 19, 1954 (R. M. Bohart); ♂, Reno, Washoe County, July 15, 1962 (F. D. Parker); ♂, Sandy, Clark County, July 24, 1958 (R. C. Bechtel); ♂, same locality, July 30, 1959 (F. D. Parker); ♀, 3 mi. SE. Schurz, Mineral County, June 27, 1961 (F. D. Parker); ♂, Smith, Lyon County, July 3, 1960 (F. D. Parker); ♂, same locality, June 26, 1961 (F. D.

Parker); 4 ♀ ♀, 9 ♂ ♂, 6 mi. NE. Sparks, Washoe County, Sept. 1, 1960, at *Chrysothamnus nauseosus* (R. W. Lauderdale); ♀, 3 ♂ ♂, same locality, Aug. 13, 1965, 4,500 ft. elev., *Chrysothamnus* sp. (H. A. Scullen); ♀, Sutcliff, Washoe County, Sept. 8, 1959 (F. D. Parker); ♀, Wadsworth, Sept. 11, 1923 (Carl D. Duncan).

PREY RECORD.—None.

PLANT RECORD.—*Chrysothamnus nauseosus; Dalea polydenia; Melilutus alba; Tetradymia canescens* (all in Nevada).

11b. *Eucerceris elegans monoensis*, new subspecies

FIGURES 14, 69 a, b, c, d, e, f

FEMALE.—Length 12 mm. Head ferruginous with limited black marks, abdomen black and yellow; punctation small and sparse; surface shiny; pubescence very short and inconspicuous.

Head slightly wider than the thorax; ferruginous except small patches bordering the eyes near the vertex, the interocellar area, antennal scrobes, apices of the mandibles, mandibular denticles, clypeal denticles and the apical part of the antennae, all of which are dark fuscous to black; clypeal border with a broad extension on the medial lobe slightly emarginate, a pair of denticles at the junction of the lobes of the clypeus, the lateral one low and rounded, a row of scattered bristles above the medial extension on the medial lobe of the clypeus; mandibles with a single bicuspidate denticle; antennae normal in form.

Thorax black except the pronotum, most of the scutellum, the metanotum, two oval patches on the enclosure, large oval patches on the propodeum, an irregular area on the pleuron and a spot on on the tegulae, all of which are yellow; the pronotum has a slight elevation on the dorso-lateral area; enclosure with a medial groove, ridged at 45° to the meson; mesosternal tubercles small; legs all ferruginous; wings subhyaline with the anterior part clouded with fulvous.

Abdomen black except for broad bands on terga 1 to 5, sterna 3 and 4, and large lateral patches on sternum 2, all of which are yellow; pygidium with sides converging to a narrow rounded apex.

MALE.—Length 12 mm. Black with yellow and ferruginous markings; punctation average; pubescence very short and inconspicuous.

Head subequal in width to the thorax; black except for most of the face, irregular spots back of the eyes, base of mandibles and small patch on the scape, all of which are yellow, and the genal area which is ferruginous; small black areas appear above the antennal scrobes; three low denticles on the medial clypeal margin, close above and between the denticles are two clusters of short bristles; mandibles without denticles; antennae normal in form.

Thorax black except the pronotum, inverted C-shaped patches on the mesoscutum, two large fused patches on the pleuron, a wide band on the scutellum, the metanotum, large patches on the propodeum and the tegulae, all of which are yellow; tegulae low and smooth; enclosure with a medial groove and ridged at about 45° to the meson; mesosternal tubercles absent; legs ferruginous except for limited elongate patches of yellow on the first and second pair of tibiae and femora; wings subhyaline but clouded along the anterior half with a darker area apically.

Abdomen black except for broad bands on terga 1 to 6, sternum 3, and variable lateral patches on sterna 2, 4, and 5; prominent rows of bristles on sterna 3 and 4, a small cluster of short bristles on sternum 5; pygidium as illustrated (fig. 69f).

Figure 14. Western U.S. *E. elegans monoensis*, new subspecies

Yellow oval patches on the enclosure are quite variable in size and may disappear from some specimens. The yellow patches on the propodeum also vary and may be divided into anterior and posterior parts. The yellow marks back of the eyes may be reduced or completely disappear.

TYPES.—The holotype ♀ and allotype ♂ are from Grant Lake, Mono County, Calif., Aug. 5, 1948 (P. D. Hurd and J. W. MacSwain). Deposited at the California Academy of Sciences.

PARATYPES.—CALIFORNIA: 9♀♀ 94♂♂, Grant Lake, Mono County, Aug. 5, 1948 (P. D. Hurd and J. W. MacSwain); 2♀♀, 6♂♂, same locality, Aug. 3, 1950, Aug. 28, 1959 (J. W. MacSwain); ♂, Lone Pine, Inyo County, June 6, 1937 (N. W. Frazier); 5♂♂, Mono Lake, Mono County, Aug. 2, 1947 (U. N. Lanham); ♀, 4 mi. S. junction Hwy. 120 and 395, Mono County, Aug. 7, 1958 (A. D. Telford).

DISTRIBUTION.—Mono County, Calif. and adjoining areas.

PREY RECORD.—None.

PLANT RECORD.—*Chrysothamnus* (California).

12. *Eucerceris ferruginosa* Scullen

FIGURES 15, 70 a, b, c

Eucerceris ferruginosa Scullen, 1939, pp. 19, 45–46, figs. 29, 69, 102, 149; 1948, pp. 159, 178; 1951, p. 1011.

TYPE.—The holotype of *E. ferruginosa* Scullen is in the California Academy of Sciences. It was taken at Angeles Bay, Gulf of California, June 26, 1921 (E. P. Van Duzee).

DISTRIBUTION.—Since the description of this species from three specimens in 1939 (p. 45) a limited number of additional specimens have come to the writer's attention. All of these are from the southwestern desert areas, mostly from California and Baja California, Mexico, except for the two specimens recorded from near Winnemucca, Nev.

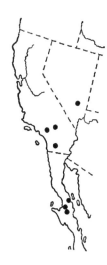

Figure 15. Southwestern U.S. and Baja California. *E. ferruginosa* Scullen

CALIFORNIA: ♀, Borrego, San Diego County, Apr. 24, 1954 (J. G. Rosen); ♀, same locality, Apr. 25, 1954, at *Croton californicus* (P. D. Hurd); 2♀♀, Brown, Inyo County, June 4, 1939, at *Eriogonum mohavense* (R. M. Bohart); 2♀♀, Llano, July 8, 1956 (Simonds); ♀, Mojave desert north of Palmdale, June 22, 1931 (F. E. Lutz); 6♀♀, 7 mi. east Mojave, Kern County, Sept. 30, 1963 (R. L. Westcott); 4♀♀, same locality and date (L. Stange); ♀, Yermo, San Bernardino County, May 31, 1941 (J. Wilcox); ♀, same locality, Sept. 7, 1952, at *Cleonella obtusifolia* (R. Snelling); NEVADA: ♀, Alamo, Lincoln County, June 26, 1958 (F. D. Parker); 2♀♀, 6. mi N., Winnemucca, Humboldt County, June 20, 1962 (R. W. Lauderdale).

MEXICO: 2♀♀, Angeles Bay, Gulf of California, June 26, 1921 (E. P. Van Duzee); ♀, 28 mi. S., El Areo, Rancho Santa Marguerita, Baja California, July 3, 1960 (A. E. Mickelbacher); 6♀♀, 20 mi. N., Mesquital, Baja California, Sept. 27, 1941 (Ross and Bohart).

The male of *E. ferruginosa* Scullen is still unrecognized. On Sept. 30, 1963 R. L. Westcott took a series of six females of this species 7 mi. E.

of the town of Mojave, Kern County, Calif. At the same time and place he took a series of five males of an undescribed species. These are being described in this paper under the name of *E. mojavensis.* Further collections and field observations may show this species to be the male of *E. ferruginosa* Scullen.

PREY RECORDS.—None.

PLANT RECORDS.—*Cleonella obtusifolia* (California). *Croton californicus* (California). *Eriogonum mohavense* (California).

13. *Eucerceris flavocincta* Cresson

FIGURES 16, 71 a, b, c, d, e, f

Eucerceris flavocinctus Cresson, 1865, pp. 109–110.—Packard, 1866, pp. 58–59.—
    Cresson, 1882, pp. vi, vii; 1887, p. 281.—Ashmead, 1890, p. 32; 1899, p.
    295.—Cresson, 1916, p. 100.—Scullen, 1939, pp. 12–14, 17, 19, 23–25, figs.
    1–13, 18, 39, 59, 77, 92, 110, 124, 139, 157, 158a–158e; 1948, pp. 155, 158,
    163; 1951, p. 1011.—Linsley and MacSwain, 1954, p. 11.—Bohart and Powell,
    1956, pp. 143–144.—Krombein, 1958, p. 197; 1960a, pp. 78–79.
Eucerceris cingulatus Cresson, 1865, pp. 110–111.—Packard, 1866, pp. 58–59.—
    Patton, 1880, p. 400.—Cresson, 1916, p. 99.
Cerceris cingulatus Schletterer, 1887, p. 488.
Cerceris flavocinctus Schletterer, 1887, p. 492.—Dalla Torre, C. G., 1897, pp.
    460–461.
Eucerceris striareata Viereck and Cockerell, 1904, pp. 84, 85, 86.—Cresson, 1928,
    p. 50.
Eucerceris chapmanae Viereck and Cockerell, 1904, pp. 84, 85, 86.—Cresson,
    1928, p. 48.

TYPES.—The holotype female of *E. flavocincta* Cresson is at the Philadelphia Academy of Natural Sciences, no. 1963. It is recorded from the Rocky Mountains, Colorado Territory (Riding). The type male of *E. cingulatus* Cresson from Colorado is at the Philadelphia Academy of Natural Sciences, no. 1964. The type female of *E. chapmanae* Viereck and Cockerell from White Oaks, N. Mex. is at the Philadelphia Academy of Natural Sciences, no. 10395.

DISTRIBUTION.—Numerous records since 1939 have extended the known range of this species further N. into parts of Canada and S. into Mexico. A male recorded from "Sierra Madre Mts., 9,700 feet" is probably from Mexico. This would be the most southern known record and the only one from Mexico. The collector and date are not given on the label.

Other collection records of special interest are the following:

CANADA: ALBERTA: ♀, 8♂♂, Lethbridge; ♀, Manyberries; ♂, Moyie; ♂, St. Mary River; ♂, Welling. BRITISH COLUMBIA: ♀, Carbonate; ♀, ♂, Chilcotin; ♀, Chopaka; 4 ♂♂, Columbia Lake; 2♂♂, Dep. Bay; 9 ♀ ♀, 4♂♂, Galiano; ♂, Goldstream; ♂, Invormere; 2 ♀ ♀ 4♂♂ Kamloops; 6 ♀ ♀, 18 ♂♂, Kaslo; 2♂♂,

Lytton; ♀, Minnie Lake; ♀, Nicola; ♂, Oliver; 2♀ ♀, 14♂ ♂, Robson, ♀, Saanich; 3♂ ♂, Salmon Arm; 2♂ ♂, Sidney; ♂, Soda Creek; ♀, Vancouver Island; ♀, ♂, Vernon; 2♀ ♀, ♂. Victoria; 2♀ ♀, Wycliffe. MANITOBA: ♂, Wawanesa; 2♂ ♂, Yale. SASKATCHEWAN: ♂, Indian Head.

UNITED STATES: ARIZONA: ♂, Grand Canyon, Bright Angel Point; ♀, Grand Canyon Swamp Lake; ♀, 3 ♂ ♂, Jacobs Lake; ♀, Kaibab Nat. Forest. MONTANA: 2♀ ♀, 4♂ ♂, Helena; ♂, Lolo Nat. Forest; ♂, Moose Lake, Ravoli County; ♀, P. & O. Ranch, Beaverhead County; ♀, Weeksville; ♀, Yaak River, Kootenai Nat. Forest, 2,612 ft. NEW MEXICO: ♂, Cimarron Canyon, Colfax Co.; 3♀ ♀, Jemez Spgs. NEVADA: ♂, Angle Lake, 8,200 ft.; ♀, ♂, Mt. Montgomery, 7,000 ft.; 2♂ ♂, Mustang, Washoe County. SOUTH DAKOTA: ♀, Black Hills Nat. Forest; 3♀ ♀, ♂, Custer Co.; ♂, Harney; ♂, Lead, Lawrence County 5,200 ft.; ♂, Pringle; 2♀ ♀, Roubaix, Lawrence County.

Figure 16. Western U.S. *E. flavocincta* Cresson

Many additional collections are recorded within the previously known range. This species is known mostly from elevations above 2,000 feet. Specimens from the northern Rocky Mountains are inclined to have the light markings more creamy white than those farther west.

PREY RECORDS.—*Dyslobus lecontei* Casey (Oregon, Scullen 1939, p. 13). *D. segnis* La Conte (Oregon, Scullen, 1939, p. 12). Bohart and Powell (1956) recorded a colony of *E. flavocincta* Cr. Independence Lake, Sierra County, Calif. which was using weevils "of an undescribed genus near *Dyslobus.*"

PLANT RECORDS.—*Achillea* (Idaho, Wyoming). *Ceanothus* sp. (California, Utah). *Conium maculatum* (Utah). *Eriogonum* sp. (California). *E. ovalifolium* (Utah). *Melilotus alba* (Oregon). *Prunus melanocampo* (Colorado). *Solidago* sp. (Oregon, Utah).

14. *Eucerceris fulvipes* Cresson

FIGURES 17, 72 a, b, c, d, e, f, g

Eucerceris fulvipes Cresson, 1865, pp. 111–112, ♀, ♂—Patton, 1880, p. 398.—
Packard, 1866, pp. 58–59.—Cresson, 1879, p. xxiii; 1882, pp. vi, vii; 1887,
p. 281.—Ashmead, 1890, p. 32; 1899, p. 295.—Smith, H. S., 1908, pp. 371,
372.—Stevens, 1917, p. 422.—Mickel, 1917, pp. 454, 456.—Viereck and
Cockerell, 1904, pp. 84, 85, 88.—Cresson, 1916, p. 100.—Pate, 1937, p. 27.—
Scullen, 1939, pp. 18, 19, 28–30, figs. 19, 50, 61, 79, 94, 112, 126, 141; 1948,
pp. 156, 158, 168; 1951, p. 1012; 1959, p. 108.
Eucerceris flavipes Ashmead, 1899, p. 295. (Apparently a mistake in spelling.)
Eucerceris simulatrix Viereck and Cockerell, 1904, pp. 84, 85, 87.—Cresson, 1928,
p. 50.—Scullen, 1959, p. 108.
Cerceris fulvipes Patton, 1879, pp. 360–361.
Cerceris cressoni Schletterer, 1887, p. 489.—Dalla Torre, C. G., 1897, p. 456.

Figure 17. Western U.S. *E. fulvipes* Cresson

TYPE.—The lectotype female of *E. fulvipes* Cresson is at the
Philadelphia Academy of Natural Sciences. It is from the Rocky
Mts., Colorado Territory (Ridings), no. 1966.1. The type female
of *E. simulatrix* Viereck and Cockerell from White Oaks, N. Mex.
is at the Philadelphia Academy of Natural Sciences, no. 1039.6.

DISTRIBUTION.—*E. fulvipes* Cresson is largely a Rocky Mts. species.
Its range extends from southern Alberta, Canada south to near the
Mexican border. Numerous females have been taken as far north as
Alberta at Medicine Hat, Manyberries and Lethbridge.

PREY RECORD.—None.

PLANT RECORDS.—*Achillea* sp. (Idaho). *Chrysothamnus* sp. (Idaho).
Cleome serrulata (Nebraska, Wyoming). *Helianthus* sp. (Nebraska).
Melilotus sp. (Nebraska). *Perideridia gairdneri* (Wyoming). *Solidago*
sp. (California, Utah). *S. canadensis* (North Dakota).

15. *Eucerceris insignis* Provancher

FIGURES 18, 73 a, b, c, d, e, f

Eucerceris insignis Provancher, 1889, p. 418.—Ashmead, 1899, p. 295.—Viereck,
1902, p. 731.—Gahan and Rohwer, 1917, p. 389.— Scullen,1939, pp. 18,
19, 43–45, figs. 28, 43, 68, 82, 101, 117, 131, 148; 1948, pp. 156, 158, 178;
1951, p. 1012.
Cerceris provancheri Dalla Torre, K. W. von, 1890, p. 204.—Dalla Torre, C. G.,
1897, p. 470.

TYPE.—The lectotype male of *E. insignis* Provancher is at the
United States National Museum. Provancher records it as a female
by mistake. It was taken in Los Angeles County, Calif. (Coquillette).

DISTRIBUTION.—California, Utah and Baja California, Mexico
with one questionable record from Utah. The writer has identified

Figure 18. Southwestern U.S. *E. insignis* Provancher

491 specimens from California, 173 specimens from Nevada (Douglas,
Ormsby, and Washoe counties) and three from Baja California
(Descanso, Ensenada and 10 mi. W. of Meling Ranch). The question-
able record from Utah is Fish Springs.

PREY RECORD.—None.

PLANT RECORDS.—(All from California) *Achillea millefolium;*
Angelica sp. *Asclepias* sp.; *Baccharis douglasii; Carum Kelloggii;*
Clematis sp.; *Daucus carota; Eriogonum fasciculatum polifolium;*
E. nudum; Eriastrum eremicum; Mentha Pulegium; Solidago sp.;
Symphoricarpos albus; Wild buckwheat.

16a. *Eucercegris lacunosa lacunosa* Scullen

FIGURES 19, 75 a, b, c, d, e, f, g, h, i, j

Eucerceris lacunosa Scullen, 1939, pp. 18, 19–20, figs. 37, 37a, 76, 109, 123, 136,
♂; 1948, pp. 155, 159; 1951, p. 1012; 1961, pp. 48–49.
Eucerceris arizonensis Scullen, 1939, pp. 18, 20–21, figs. 14, 57, 90, 137, ♀ ; 1948,
p. 157; 1951, p. 1011.

TYPES: The type male of *E. lacunosa* Scullen from Bill Williams
Fork, Ariz., August (F. H. Snow) is in the collection of the University

of Kansas. The type female of *E. arizonensis* Scullen from Oslar, Huachuca Mts., Ariz., is also in the collection of the University of Kansas.

Many specimens of this species have been recorded by the writer since its description in 1939. North of the Mexican border it appears to be confined to southern Arizona, southern New Mexico and Southwestern Texas. It has been taken S. through the state of Chihuahua and in the state of Durango, Mexico.

DISTRIBUTION.—Southern Arizona, southern New Mexico, southwestern Texas, and north central Mexico. Specimens are as follows (collectors and dates are omitted for brevity):

ARIZONA: ♀, southern Arizona; ♀, Apache, Cochise County; ♀, 3 ♂♂, Skeleton Canyon, 6 mi. E. Apache, Cochise County; ♂, Bill Williams Fork; 2 ♂ ♂, 35 mi. E. Douglas; ♂, Dragoon, Cochise County; 2 ♂ ♂, Higley, Maricopa

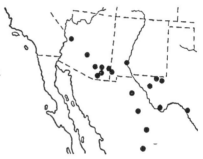

Figure 19. Southwestern U.S. and western Mexico. *E. lacunosa lacunosa* Scullen

County; 8 ♀ ♀, ♂, Huachuca Mts., Cochise County; 2 ♂ ♂, 12 mi. N. Huachuca, Cochise County; 19 ♀ ♀, 105 ♂ ♂, Portal area, Cochise County; ♂, San Simon, Cochise County; ♀, Tombstone, Cochise County; ♀, ♂, Tucson, Pima County; ♂, 20 mi. W. Tucson; ♂, 18 mi. SW. Tucson; ♀, 4 ♂ ♂, Willcox, Cochise County.

NEW MEXICO: ♂, 25 mi. N. Las Cruces, Dona Ana County; ♀, 21 ♂ ♂, Rodeo, Hidalgo County. TEXAS: 2 ♀ ♀, Chisos Mts., Big Bend Nat. Park; ♂, Eagle Pass, Maverick County; ♂, Glenn Springs, Brewster County; ♀, 10 mi. N. Pecos, Reeves County; ♀, 4 mi. S. Pecos, Reeves County; ♀, 20 mi. E.-NE. Pine Springs, Culberson County; 2 ♀ ♀, ♂, Van Horn, Culberson County. MEXICO: 2 ♀ ♀, Chihuahua, Chih.; ♀, Gomez Palacio, Dgo.; ♀, 6 ♂ ♂, 18 mi. W. Jimenez, Chih.; ♀, Moctezuma, Chih.; ♀, 15 mi. W. Verdo, Nazas River, Dgo.

PREY RECORDS.—None.

PLANT RECORDS.—*Asclepias* sp. (New Mexico). *Baccharis glutinosa* (Arizona, Chihuahua, Mexico). *Cleome* sp. (Arizona). *Eriogonum abertianum neomexicanum* (Arizona). *Haplopappus* (=*Aplopappus*) *hartwegi* (Arizona). *Koeberlinia spinosa* (New Mexico). *Melilotus alba* (Portal, Ariz.). *Mortonia scabrella* (Arizona). *Robinia* sp. (Texas).

16b. *Eucerceris lacunosa sabinasae*, new subspecies

FIGURE 20

MALE.—The male of this subspecies agrees with the male of the nominate subspecies in all respects except for the following color markings.

Head is the same except *E. lacunosa sabinasae* has no yellow back of the eye.

Thorax of *sabinasae* has only a trace of yellow on the posterior margin of the pronotum and the metanotum.

Abdomen of *sabinasae* has the yellow bands of the terga deeply emarginate with ferruginous on all but tergum 6 and the venter is immaculate.

FEMALE.—The female of this subspecies agrees with the female

Figure 20. Northern Mexico. *E. lacunosa sabinasae*, new subspecies

of the nominate subspecies in all respects except for the following color markings.

The head of *sabinasae* is ferruginous except for a narrow yellow line bordering the eye on the face.

The thorax is ferruginous except for a trace of yellow on the metanotum, and black along most of the sutures.

The abdomen has yellow bands reduced to posterior ridges of terga 2, 3, and 4; tergum 5 has traces of yellow; the venter is immaculate.

TYPES.—The ♂ type of *Eucerceris lacunosa sabinasae* Scullen is from 23 mi. N. of Sabinas, Coah., Mexico, Aug. 10, 1959 (A. S. Menke and L. A. Stange). The allotype ♀ is from Quemado, Maverick County, Tex., Sept. 3, 1960 (L. A. Stange). These are deposited with the University of California at Davis.

PARATYPES.—ARIZONA: ♂, 2 mi. NE. Portal, Cochise County, Aug. 17, 1962 (H. A. Scullen). MEXICO: 4♂♂, 23 mi. N. Sabinas, Coah., Mex., Aug. 10, 1959 (A. S. Menke and L. A. Stange).

DISTRIBUTION.—Northern Coahuila, Mex., Maverick County, Tex. and Cochise County, Ariz.

PREY RECORDS.—None.

PLANT RECORDS.—*Baccharis glutinosa* (Arizona).

17. *Eucerceris lapazensis*, new species

FIGURES 21, 74 a, b, c

FEMALE.—Length 15 mm. Ferruginous with yellow markings and some depressed areas darker; punctation small and mostly crowded; pubescence very short.

Head slightly wider than the thorax, ferruginous becoming more fulvous over the lower face, a narrow wedge of fuscous borders the eye near the vertex, a depressed fuscous area extends through the antennal scrobes and along the sutures of the lower face; clypeal border with two pronounced concavities on the medial lobe, laterad

Figure 21. Baja California. *E. lapazensis*, new species

of each cavity there is a low denticle with a broad base, at the meson between the two depressions there is a cluster of bristles, below these bristles there is a pair of denticles connected with each other by a carina, another similar carina connects the medial denticles with each lateral denticle; mandibles with a single acute denticle; antennae normal in form, basal segments ferruginous, apical segments darker.

Thorax dark ferruginous except for the posterior border of the pronotum, its posterior lobe, the metanotum and an area on the propodeum, all of which are more yellow; tegulae low and smooth; enclosure with a deep medial groove and ridged subparallel to the base; mesosternal tubercles small; legs ferruginous; wings subhyaline but clouded in the anterior area, second submarginal cell petiolate.

Abdomen ferruginous except for a broad emarginate band on tergum 1, wide but deeply and broadly emarginate bands on terga 2, 3, and 4, a broad band with an elongate narrow patch of ferruginous on tergum 5, and irregular patches on sterna 2 and 3, all of which are yellow infused with ferruginous; pygidium with sides slightly convex tapering to a small rounded apex.

MALE.—Unknown.

TYPE.—The ♀ type is from La Paz, Baja California, Mexico, Oct. 12, 1954 (F. X. Williams). Deposited at the California Academy of Sciences.

PARATYPES.—MEXICO: ♀ Santo Domingo, Baja California, Nov. 16, 1941 (Frank Gander); ♀, "Basse-Californie Diguet 178–95."

DISTRIBUTION.—Known only from Baja California, Mex.

PREY RECORDS.—None.

PLANT RECORDS.—None.

18. *Eucerceris melanosa* Scullen

FIGURES 22, 76 a, b, c

Eucerceris melanosa Scullen, 1948, pp. 156, 163, figs. 3a, b, c, 13; ♂.

MALE.—The male was described in 1948 from a single specimen. A limited number of specimens has been taken since then.

FEMALE.—The female of *E. melanosa* Scullen has never been recognized. *E. menkei* Scullen, being described in this paper, has color markings similar to *E. melanosa* Scullen and may prove to be the female of that species.

TYPE.—The holotype male of *E. melanosa* Scullen is at the University of Minnesota. It was taken at Tehuacan, Pueb., Mexico, July 12, 1935 (A. E. Pritchard).

DISTRIBUTION.—This species was described (Scullen, 1948, p. 163) from the unique male. Since that time eight additional males have been recorded as listed below. The female has not yet been recognized. Specimens are as follows:

MEXICO: ♂, Calcaloapan, Pueb., Aug. 26, 1962, (F. D. Parker); 2♂ ♂, same locality, Aug. 20, 1963, (Parker and Stange); ♂, Mexico, D. F., Aug. 26, 1928 (G. Lassmanu); ♂, 38 mi. SE., Oaxaca, Oax., 5,600 ft. elev., Aug. 19, 1963 (Scullen and Bolinger); 3♂ ♂, 4 and 5 mi. W. of Pachuca, Hdgo., 7,900 ft. elev., June 16, 1961, Aug. 25, 1962, (Roberts and Nauman); ♂, 41 mi. N. Queretaro, Qro., 6,500 ft. elev., Sept. 19, 1963 (Scullen and Bolinger); 2♂ ♂, 18 mi. W., Tehuacan, Pueb., 6,200 feet elev., Sept. 5, 1957 (H. A. Scullen).

PREY RECORDS.—None.

PLANT RECORDS.—None.

19. *Eucerceris melanovittata* Scullen

FIGURES 23, 77 a, b, c, d, e, f, g

Eucerceris melanovittata Scullen, 1948, pp. 156, 164, figs. 4a, b, c, 14; 1951, p.1012.

This species was described by the writer in 1948 (p. 164) from seven male specimens. Since that time 41 additional males have been determined. In addition to the above males two females have recently been recognized. The female is described below:

FEMALE.—Length 15 mm. Black and ferruginous with light yellow markings; punctation and pubescence average.

Head subequal in width to the thorax; black and ferruginous except large frontal eye patches, large patches on the lateral and medial clypeal lobes, large patches on the genae, small evanescent

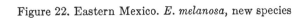

Figure 22. Eastern Mexico. *E. melanosa*, new species

spots on the vertex and lateral spots on the base of the mandibles, all of which are light yellow; the following parts are ferruginous: lower part of the medial lobe of the clypeus, most of the genae exclusive of the yellow patch, the basal half of the mandibles, and the basal third of the antennae; the black vittae of the face extend through the antennal scrobes to the ventral border of the clypeus; mandibles with one medial denticle; antennae normal in form.

Thorax black and ferruginous except a band on the pronotum extending onto the posterior lobe, a broad band deeply emarginate on the scutellum, the metanotum, large patches on the propodeum, a patch on the pleuron, four small patches on the sternum and large spots on the tegulae all of which are light yellow; ferruginous areas include most of the pleura, most of the sternum, the surrounding border of the yellow patches on the propodeum and most of the tegulae; tegulae normal in form; enclosure with a medial groove and light ridges extending laterally from the medial groove; mesosternal tubercles absent; legs ferruginous; wings subhyaline except for a darker area apically along the anterior border, second submarginal cell petiolate.

Abdomen black with limited ferruginous except for broad bands more or less emarginate on terga 1 to 5 and sterna 3 and 4, and small lateral spots on sternum 2, all of which are light yellow; ferruginous largely replaces the black on the 1st tergum, the 1st and 2nd sterna and the entire 6th segment; pygidium with sides slightly convex converging to a rounded apex.

Two specimens of females of *E. melanovittata* Scullen have been recognized. The above description is based largely on one taken at Willcox,

Figure 23. Southwestern U.S. *E. melanovittata* Scullen

Cochise County, Ariz., July 7, 1956 (A. D. Telford). Another specimen was taken at Davis Mts., Jeff Davis County, Tex., Sept. 4, 1949 (F. Werner and W. Nutting). The latter specimen has markings much lighter than the former and the ferruginous area is very limited on the body parts. Males taken from each of the above localities show similar color variations. Most of the males examined show no ferruginous. Future collections may show it desirable to recognize two subspecies based on the extent of the ferruginous markings.

Two males taken 40 mi. W. of Linares, N. L. Mexico, have the legs very dark and the body parts also darker than the typical form. Further collections may show it desirable to recognize this color form as a distinct subspecies also.

TYPE.—The holotype male of *E. melanovittata* Scullen is at the California Academy of Sciences. It was taken 25 mi. E. of El Paso, Tex., July 13, 1942 (E. C. Van Dyke).

DISTRIBUTION.—Southern Arizona, central and southern New Mexico and western Texas with one questionable record each from western Nevada and Sioux County, Nebr.; Nuevo Leon, Mexico. The following is a complete record of all specimens determined to date including previously published records:

ARIZONA: ♂, 10 mi. E. Douglas, Cochise County, July 11, 1940 (E. S. Ross); ♂, Portal, Cochise County, Sept. 18, 1958 (H. V. Weems, Jr.); ♂, 2 mi. NE. Portal, Cochise County, Sept. 3, 1960, at *Baileya multiradiata* (M. Cazier); 2♂♂, same locality, 4,700 ft. elev., May 30, 1961, Aug. 14, 1962, at *Thelesperma megapotamicum, Baccharis glutinosa* (H. A. Scullen); ♂, 2 mi. SE. Portal, Cochise County, July 2, 1960 (M. Cazier); ♂, 5 mi. W. Portal, Cochise County, 5,400 ft. elev., July 18, 1956, at *Melilotus alba* (M. Cazier); ♂, 24 mi. S. Stafford, Graham County, 4,400 ft. elev., Sept. 14, 1962, at *Eriogonum* sp. (H. A. Scullen); ♀, Willcox, Cochise County, July 7, 1956 (A. D. Telford); ♂, same locality, Aug. 18, 1958 (R. M. Bohart); ♂, 3 mi. SE. Willcox, Cochise County, Aug. 29, 1957, at *Cleome* sp. (W. F. Barr). NEBRASKA: ♂, 7 mi. N. Harrison, Sioux County, Aug. 13, 1962, at *Helianthus* (J. G. and B. L. Rozen). NEVADA: ♂, Ormsby County, July 6, (Baker). NEW MEXICO: 11♂♂, Duran, Torrance County, July 13, 1959 (E. G. Linsley); ♂, Jemez Springs, Sandoval County, Aug. 15, 1957 (D. J. and J. N. Knull); ♂, Los Montoyas, San Miguel County, July 13, 1959, at *Melilotus alba* (E. G. Linsley); ♂, Mountainair, Torrance County, 1924 (C. H. Hicks); ♂, Santa Fe, July 4, 1934 (F. E. Lutz); ♂, same locality, Sept. 2, 1934 (P. E. Geier). TEXAS: 2♂♂, 10 mi. S. Alpine, Brewster County, 5,200 ft. elev., Sept. 1 and 7, 1962 (H. A. Scullen); ♂, Big Bend Nat. Park, Brewster County, Aug. 31, 1962 (H. A. Scullen); ♂, Chisos Mts., Big Bend Nat. Park, Brewster County, July 6, 1942 (H. A. Scullen); ♀, Davis Mts., Jeff Davis County, Sept. 4, 1949 (F. Werner and W. Nutting); ♂, 25 mi. E. El Paso, July 13, 1942 (E. C. Van Dyke); ♂, Ft. Davis, Jeff Davis County, Sept. 6-7, 1943 (R. W. Strandtmann); ♂, Guadalupe Pass, Hudspeth County, July 28, 1950 (Ray F. Smith); 2♂♂, W. side Hueco Mts., El Paso County, 4,570 ft. elev., Aug. 30, 1937 (Rehn, Pate, Rehn); ♂, Marathon, Brewster County, July 7, 1942 (E. C. Van Dyke); 3♂♂, 8 mi. E. Marfa, Presidio County, 4,650 ft. elev., Sept. 2, 1962, at *Eriogonum annum* (H. A. Scullen); ♂, Santa Elena Canyon, Big Bend Nat. Park, Brewster County, 2,145 ft. elev., Aug. 25, 1954 (R. M. Bohart); ♂, Sierra Blanca, Hudspeth County, July 8, 1917 (Jos Bequaert). MEXICO: 2♂♂, 40 mi. W. Linares, N. L., 5,200 ft. elev., Sept. 7, 1963 (Scullen and Bolinger).

PREY RECORDS.—None.

PLANT RECORDS.—*Baccharis glutinosa* (Arizona). *Baileya multiradiata* (Arizona). *Cleome* sp. (Arizona). *Eriogonum* sp. (Arizona). *E. annum* (Texas). *Helianthus* (Nebraska). *Melilotus alba* (Arizona, New Mexico). *Thelesperma megapotamicum* (Arizona).

20. Eucerceris mellea Scullen

FIGURES 24, 78 a, b, c, d, e, f

Eucerceris mellea Scullen, 1948, pp. 156, 158, 165–166, figs. 5a,b,c,d,e,f, 14; 1951, p. 1012.

TYPE.—The holotype female and allotype male of *E. mellea* Scullen are at the California Academy of Sciences. Both were taken in the

Chisos Mts., Big Bend Nat. Park, Brewster County, Tex., July 6, 1942 (H. A. Scullen).

DISTRIBUTION.—Since this species was described by this writer in 1948 (p. 165) a few new records have shown an enlarged known distribution including Chihuahua, Mexico. However, only the one female has been recorded. The types and paratypes are included in the following records. Specimens are as follows:

NEW MEXICO: Rowe, San Miguel County, July 19, 1952 (R. H. Beamer and party). TEXAS: ♀, 45♂♂, Chisos Mts., Big Bend Nat. Park, Brewster County, July 3–6, 1942 (H. A. Scullen); 8♂♂, same locality, July 6, 1942 (E. C. Van Dyke); ♂, same locality, July 17, 1921 (Carl D. Duncan); ♂, same locality (no date or collector); ♂, same locality (J. Bequaert); ♂, Davis Mts., Jeff Davis County, June 26, 1942 (H. A. Scullen); ♂, 11 mi. N., 4 mi. W., Alpine, June 20, 1961 (D. W. Smith). MEXICO: 2♂♂, Chihuahua, Chih., Aug. 12, 1951, at *Baccharis glutinosa* (P. D. Hurd).

PREY RECORDS.—None.

PLANT RECORDS.—*Baccharis glutinosa* (Chihuahua, Mexico)

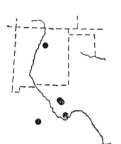

Figure 24. Rio Grande Valley. *E. mellea* Scullen

21. *Eucerceris menkei*, new species

FIGURES 25, 79 a, b, c, d

FEMALE.—Length 13 mm. Black with yellow markings becoming fulvous on the distal half of the legs; punctation small and crowded; pubescence longer than average on many parts.

Head slightly wider than the thorax; black except for frontal eye patches, patches on the medial clypeal lobe, and each lateral lobe, an elongate area between the antennal scrobes and an elongate patch back of the eye, all of which are dusky yellow; clypeal border with two pair of denticles, each pair located near the junction of the medial and lateral lobes of the clypeus; the two medial pair of denticles connected with each other by an emarginate lamella; medially and just above this lamella there is a small cluster of bristles; mandibles with two denticles, one above the other, the dorsal one much the larger and apically rounded; antennae normal in form, largely fulvous basally but darker apically.

Thorax black except for a line on the pronotum, the metanotum, a patch on the propodeum and a small spot on the pleuron, all of which are yellow; tegulae low and smooth; enclosure with a deep medial groove and heavily ridged subparallel to the base; mesosternal tubercles absent; legs dark fuscous basally to near the apical end of the femori beyond which they are fulvous; wings subhyaline but clouded with fulvous, darker along the anterior part, second submarginal cell petiolate.

Abdomen black except for a deeply emarginate band on tergum 1, a narrow posterior band on tergum 2, a broken anterior line on tergum 3, a divided band on terga 4 and 5, tergum 6 exclusive of the pygidium, and broad but emarginate semidivided bands on sterna 3, 4, and 5, all of which are yellow; pygidium with sides slightly convex converging to a narrow rounded apex.

MALE.—Unknown.

Figure 25. Mexican Gulf Coast. *E. menkei,* new species

E. menkei Scullen is approximately the same size and has about the same color pattern as *E. stangei* Scullen being described from the state of Oaxaca. The clypeal structures of the females are very different. The very unusual position of the mandibular denticles shows a close relationship to *E. sinuata* Scullen. *E. menkei* Scullen is known only from the unique female which is very different from any other known species. The color pattern would indicate it could prove to be the female of *E. melanosa* Scullen.

TYPE.—The holotype female of *E. menkei* Scullen was taken 10 mi. N.W. of Tamazulapam, Oax., Mexico, Aug. 22, 1959 (L. A. Stange and A. S. Menke). Deposited at the California Academy of Sciences.

DISTRIBUTION.—Known only from the type locality of Tamazulapam, Oaxaca, Mex.

PREY RECORD.—None.

PLANT RECORD.—None.

22. *Eucerceris mojavensis,* new species

FIGURES 26, 80 a, b, c

MALE.—Length 11 mm. Black with creamy yellow and ferruginous markings; punctation more limited than average; pubescence avery short.

Head subequal in width to the thorax; black except the entire face exclusive of the antennal scrobes, an elongate patch on the gena and basal two thirds of the mandibles, all of which are creamy yellow; ventral parts of the head, much of the gena, basal segments of the antennae, and clypeal border are ferruginous; clypeal border with three subequal denticles; hair lobe bristles not matted, extending somewhat onto the medial lobe; mandibles without denticles; antennae normal in form.

Figure 26. Southwestern U.S. *E. mojavensis,* new species

Thorax black except a band on the pronotum extending onto the posterior lobe and the episternum, four stripes on the mesoscutum each lateral pair fused anteriorly, a broad band on the scutellum, the metanotum, patches on the tegulae, large patches on the propodeum, a pair of wedge-shaped small patches on the enclosure, most of the mesopleuron, all of which are creamy yellow; tegulæ normal in form; mesosternal tubercles absent; legs ferruginous except for patches on all coxæ and all trochanters, elongate patches on the fore and mid femora, the fore and mid tarsi, and the apex of the hind femora, all of which are yellow; wings subhyaline with a clouded area along the anterior margin becoming darker apically.

Abdomen with broad bands of creamy yellow on terga 1 to 5; tergum 6 with a medial creamy yellow patch on a background of ferruginous; venter ferruginous with bands on sterna 2 and 3, lateral patches on sterna 4 and 5, all of which are creamy yellow; broad rows of bristles on sterna 3 and 4, a small divided cluster of short bristles on sternum 5; pygidium as shown (fig. 80c).

The series of males taken at the type location Sept. 30, 1963 was taken at the same time and place as a series of females of *E. ferruginosa* Scullen and may prove to be the unknown male of that species.

TYPE.—The ♂ type of *E. mojavensis* Scullen was taken 7 mi. E. of Mojave, Kern County, Calif., Sept. 30, 1963 (R. L. Westcott). It is at the University of California at Davis.

PARATYPES.—CALIFORNIA: 4 ♂ ♂, 7 mi. E. of Mojave, Kern County, Sept. 30, 1963 (R. L. Westcott); ♂, Lone Pine, Inyo County, June 18, 1937 (E. C. Van Dyke).

DISTRIBUTION.—Eastern Kern County and Inyo County Calif.
PREY RECORDS.—None.
PLANT RECORDS.—None.

23. *Eucerceris montana* Cresson

FIGURES 27, 28, 81 a, b, c, d, e, f, g, h, i

Eucerceris montanus Cresson, 1882, pp. vi, vii, viii; 1887, p. 281.—Ashmead, 1899, p. 295.—Viereck and Cockerell, 1904, pp. 84, 85, 86, 87.—Cresson, 1916, p. 101 —Mickel, 1917, pp. 454, 456.—Scullen, 1939, pp. 17, 18, 54–56, figs. 33, 47, 74, 88, 107, 121, 134a, 155; 1948, pp. 157, 158, 180; 1951, p. 1012; 1961, p. 49
Cerceris montana Dalla Torre, K. W. Von, 1890, p. 201.
Cerceris sonorensis Cameron, 1891, p. 129.

TYPE.—The lectotype female of *E. montana* Cresson from Montana (Morrison) is at the Philadelphia Academy of Sciences, no. 1946. The holotype male of *Cerceris sonorensis* Cameron from northern Sonora, Mex., is in the British Museum, no. 21. 1, 435.

DISTRIBUTION.—Although this species was described from one female and two males from "Montana" the writer has never seen any other specimens from that far north. The most northern records are three males from northern Utah and a female from Sterling, Colo. We have found *E. montana* Cr. common to abundant in southern Arizona, southern New Mexico and western Texas. It has been collected by the writer and others south of the border over the central plateau area of Mexico as far south as the states of Jalisco and San Luis Potosí.

PREY RECORDS.—Abundant as this species is nothing is known relative to its nesting habits or the prey used as food for the young.

PLANT RECORDS.—*Acacia* sp. (Texas). *A. gregii* (Texas). *Asclepias linaria* (Mexico: Zacatecas). *A. subverticillata* (New Mexico). *Bac-*

charis glutinosa (Arizona, Texas, Mexico: Chihuahua). *Buddleja scordioides* (Texas). *Croton corymbulosus* (Texas). *Dodder (Cuscuta* sp.) (Texas). *Eriogonum* sp. (Arizona). *E. annuum* (Texas). *Eysenhardtia polystachya* (Mexico: Jaliseo). *Guardiola tulocarpa* (Mexico: Durango). *Koeberlinia* sp. (Texas). *Lepidium thurberi* (New Mexico). *Melilotus alba* (Arizona, Utah). *Prosopis juliflora* (Texas). *Sartiwellia mexicana* (New Mexico). *Solidago* sp. (Texas).

Figure 27. Western U.S. *E. montana* Cresson

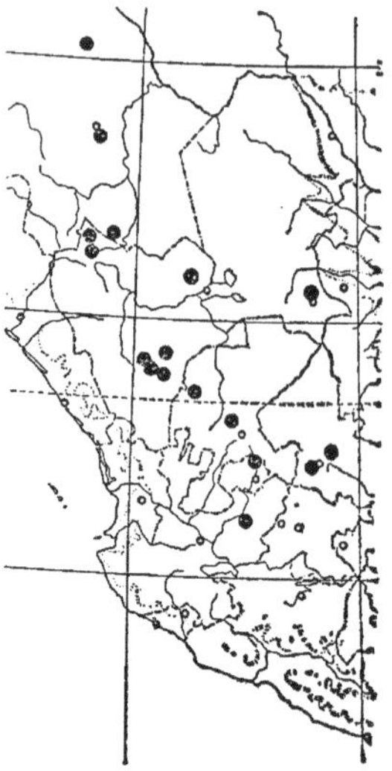

Figure 28. Western Mexico. *E. montana* Cresson

24a. *Eucerceris morula albarenae*, new subspecies

FIGURE 29

FEMALE.—Length 12 mm. Black and ferruginous with creamy white markings; punctation average; pubescence very short.

Head one seventh wider than the thorax; black except for large frontal eye patches, large patches on the medial and lateral lobes of the clypeus, a constricted patch between the antennal scrobes, an elongate patch back of the eye and a spot on the base of the mandible, all of which are creamy white; limited amounts of ferruginous border the light patches back of the eyes, cover much of the mandibles and the more basal segments of the antennae; clypeal border divided into an upper and a lower portion on the medial lobe, single low denticles are located at the fusion of the medial lobe with each lateral lobe, a pair of dentciles is at the meson of the upper portion of the border between which there is a cluster of bristles, a bilobed extension appears mesad on the lower portion of the border; mandibles with a single large acute denticle with a broad base; antennae dark apically, lighter basally, normal in form.

Thorax black except a band on the pronotum extending onto the posterior lobe, a deeply emarginate band on the scutellum, the metanotum, large single patches on the propodeum, a triangular patch on the pleuron and a spot on the tegulae, all of which are creamy white; evanescent areas of light fuscous appear on the pleuron below the light patches and at the meson of the scutellum; tegulae normal in form; enclosure with a deep medial groove and heavy lateral ridges subparallel to the base; mesosternal tubercles present but very small; legs ferruginous; wings subhyaline with a dark area on the anterior border embodying the marginal cell and the apex beyond.

Abdomen black except for broad and emarginate bands on terga 1

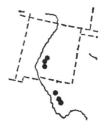

Figure 29. Rio Grande Valley. *E. morula albarenae*, new subspecies

to 5 which are creamy white and the basal parts of terga 1 and 2 which are infused with ferruginous and all sterna which are ferruginous; pygidium with sides slightly convex and converging to a rounded apex.

MALE.—Length 11.5 mm. Black with creamy white markings on the body; legs ferruginous with creamy white markings; punctation average; pubescence very short.

Head one seventh wider than the thorax; black except for the entire face exclusive of black lines through the antennal scrobes (the light area fusing above the antennae), a light spot back of the eye, basal parts of the mandibles and a minute spot on the scape, all of which are creamy white to very light yellow; clypeal border with three dark subequal denticles; mandibles without denticles; antennae normal in form.

Thorax black except for a broad band on the pronotum extending onto the posterior lobe, an elongate patch on the episternum, an emarginate band on the scutellum, the metanotum, large lateral patches and small more medial spots on the propodeum, a pair of small spots on the enclosure near the apex, a patch on the pleuron and a spot on the tegulae, all of which are creamy white; tegulae normal; enclosure with a deep medial groove and heavily ridged subparallel to the base; mesosternal tubercles absent; legs ferruginous with light yellow patches of variable sizes on all coxae, all trochanter, the fore and mid femora and the fore and mid tibiae; wings subhyaline with a dark area near the apical end of the stigma and a second dark area near the apex of the fore wing; second submarginal cell not petiolate.

Abdomen black except for a broad band on tergum 1, a broad band enclosing a dark area in the depression on tergum 2, narrow bands broadening laterally on terga 3, 4, and 5, narrow band on tergum 6, spot on sternum 1, bands somewhat emarginate on sterna 2, 3, and 4, and small lateral spots on sternum 5, all of which are creamy white; wide prominent rows of bristles are on the apical margins of sterna 3 and 4; pygidium as illustrated (fig. 29).

The subspecies *E. morula albarenae* Scullen differs from the nominate subspecies as follows:

E. morula morula Scullen	*E. morula albarenae* Scullen
Female	
Two patches on the scutellum.	Complete band on scutellum.
A divided band or two patches on tergum 1.	Emarginate band on tergum 1.
All parts of body black with creamy white marks.	Terga 1 and 2 and all sterna more or less infused with ferruginous.
Male	
Yellow of face not closed above the antennae (i.e., vittae of the face fuse with the black of the vertex).	Yellow of the face closed above the antennae (i.e., vittae of the face do not extend to the vertex). (Intermediate conditions are found.)
Evanescent spots may appear on the enclosure and small evanescent spots may appear on the propodeum near the apex of the enclosure. In rare cases these latter spots may be joined to the larger lateral patches of the propodeum giving an inverted "C" shaped patch.	No evanescent spots on the propodeum or the enclosure.
Band on tergum 1 broad and uninterrupted.	Band on tergum 1 divided or deeply emarginate.

TYPES.—The type ♀ and allotype ♂ of *E. morula albarenae* Scullen were taken 31 mi. NE. of Las Cruces, Otero County, N. Mex. (near the White Sands National Monument), Sept. 9, 1962, at 4,000 ft. elev. on *Sartwellia mexicana* (H. A. Scullen). They are at the U.S. National Museum, no. 69228.

PARATYPES.—NEW MEXICO: ♂, Eddy County, June 21, 1940 (M. Price); 8♀♀, 19♂♂, 31 mi. NE., Las Cruces, Otero County, 4,000 ft. elev., Sept. 9, 1962, at *Sartwellia mexicana* (H. A. Scullen); ♂, 35 mi. NE. Las Cruces, Otero County, 3,900 ft. elev., Aug. 21, 1962, at *Eriogonum rotundifolium* (H. A. Scullen); 3♀♀, 4♂♂, same locality, Sept. 9, 1962, at *Sartwellia mexicana* and *Lepidium* sp. (H. A. Scullen); ♂, 7.5 mi. S. Three Rivers, Otero County, Sept. 9, 1961, at *Gutierrezia micrecephala* (P. D. Hurd); 6♀♀, 58♂♂, White Sands Monument Area, Otero County, 4,000 ft., elev., Sept. 9, 1962, at *Sartwellia mexicana* (H. A. Scullen) TEXAS: 5♀♀, 38♂♂, 12 mi. W. Alpine, Brewster County, 5,000 ft., elev., Sept. 3, 1962, at *Eriogonum annum* (H. A. Scullen); ♀, 7 mi. NE. Marfa, Presidio County, 4,800 ft. elev., Sept. 2, 1962, at *Croton corymbulosus* (H. A. Scullen); 2♀♀, 4♂♂, 8 mi. E. Marfa, Presidio County, 4,650 ft. elev., Sept. 2, 3, 7, 1962, at *Eriogonum annum* and *Sartwellia mexicana* (H. A. Scullen); 4 ♂♂, 15 mi. E. Van Horn, Brewster County, Sept. 28, 1957, at *Gutierrezia sarothrae* (Nutting and Werner).

DISTRIBUTION.—Central New Mexico and western Texas, 3,900 to 5,000 ft. elev.

PREY RECORD.—None.

PLANT RECORD.—*Croton corymbulosus* (Texas). *Eriogonum annum* (Texas). *Eriogonum rotundifolium* (New Mexico). *Gutierrezia microcephala* (New Mexico). *G. sarothrae* (Texas). *Sartwellia mexicana* (New Mexico, Texas).

24b. *Eucerceris morula morula*, new species

FIGURES 30, 82 a, b, c, d, e, f

FEMALE.—Length 10 mm. Black with creamy white markings; punctation average; pubescence very short and inconspicuous.

Head about one sixth wider than the thorax; black except large eye patches, large patches on the lateral lobes and the medial lobe of the clypeus, an elongate patch on the frons, an elongate patch back of the eye, all of which are creamy white; clypeal border divided into an upper and a lower portion on the medial lobe, single denticles are located at the fusion of the medial lobe with each lateral lobe, a pair of denticles is at the meson of the upper portion of the border between which there is a group of 2 to 4 long bristles, a bilobed extension appears mesad on the lower portion of the border; the surface of the medial clypeal lobe is only slightly convex; mandibles black with a large acute denticle having a broad base, below and more basad there is a small denticle (visible only when specimen is viewed from a certain angle); antennae black and normal in form.

Thorax black except for a band on the pronotum extending onto the posterior lobe, two patches on the scutellum, the metanotum, large patches on the propodeum, a patch on the pleuron below the wing, and a spot on the tegula, all of which are creamy white; tegulae low and smooth; enclosure with a medial groove and lightly ridged at about 45°; mesosternal tubercle very small; all coxae, trochanters, and the first femur black except for a creamy white spot on the above femur, other leg parts ferruginous becoming darker on the third tarsi; wings subhyaline clouded at the apex; the second submarginal cell is petiolate.

Abdomen black except for a divided band on tergum 1, emarginate bands on terga 2, 3, 4, and 5; venter immaculate, pygidium with the sides slightly convex converging to a rounded apex.

MALE.—Length 11 mm. Black with creamy white markings; punctation average; pubescence very short and inconspicuous.

Head very slightly wider than the thorax; black except for very

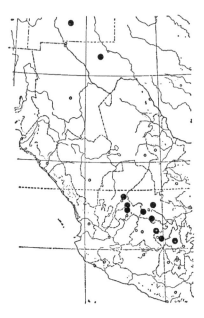

Figure 30. Central Mexico. *E. morula morula*, new species

large eye patches, the entire clypeus, basal two thirds of the mandibles and a small spot back of the eye, all of which are creamy white; the black vitta on the face extend to the clypeal border and fuse with the black of the vertex; clypeal border with three subequal black denticles on the medial lobe; bristle-like setae over the entire lower clypeal area; mandibles without denticles; antennae black and normal in form.

Thorax black except for a band on the pronotum extending onto the posterior lobe, a divided band on the scutellum, the metanotum, large patches on the propodeum, a patch on the pleuron below the wing and a spot on the tegula, all of which are creamy white; tegulae low and smooth; enclosure with a medial groove and ridged at a 45° to 60° angle with the mesal groove; mesosternal tubercle absent;

all coxae and trochanters black marked with considerable creamy white, fore femora black with large creamy white patches apically, remaining parts of all legs largely ferruginous with a large creamy white patch on the mid femora and small spots on the fore and mid pair of tibiae; wings subhyaline with dark patches near the apex; the 2nd submarginal cell is not petiolate.

Abdomen black except for a broad semidivided band on tergum 1, emarginate bands on terga 2, 3, 4, 5, and 6, bands on sterna 2 and 3, and lateral patches on sternum 4, all of which are creamy white; wide prominent rows of bristles appear inserted on the apical margins of sterna 3 and 4, a very indistinct pair of bristle clusters may appear on sternum 5; the pygidium has the form illustrated (fig. 82f).

Both sexes of *E. morula morula* Scullen closely resemble *E. vittatifrons* Cresson in size and color. The females may be distinguished by the structure of the clypeal borders and the males by the color pattern of the face, the abdominal bristles and the venation of the wings. The male of *E. morula morula* is close to the male of *E. baja* Scullen but is distinguished from that species by the ferruginous area back of the eye and a very distinct row of bristles on sternum 5 on the latter species. The male of *E. pacifica* Scullen is also very close, but the key should separate them. (See notes under *E. morula albarenae* Scullen, p. 48.)

TYPES.—The type ♀ and allotype ♂ of *E. morula morula* Scullen were collected 17 mi. NE. of San Luis Potosí, S. L. P., Mexico, at 6,200 ft. elev., Sept. 6, 1963 on *Baccharis glutinosa* (H. A. Scullen and Duis Bolinger). They are at the U.S. National Museum, No. 69229.

PARATYPES.—MEXICO: ♂, 40 mi. N. Aguascalientes, Ags., 6,600 ft. elev., June 15, 1956 (H. A. Scullen); 3 ♂ ♂, 1.5 mi. S. Fresnillo, Zac., Aug. 6, 1954, at *Solidago* (E. Linsley, J. MacSwain, and R. Smith); ♀, 6 ♂ ♂, 9 mi. S. Fresnillo, Zac., Aug. 10, 1954 (E. Linsley, J. MacSwain, and R. Smith); 3 ♂ ♂, Ixmiquilpan, Hgo., 5,300 ft. elev., June 23, 1953, at *Sphaeralcea* (U. of Kansas Mex. Exped.); 3 ♀ ♀, 7 ♂ ♂, 41 mi. N. Queretaro, Qro., 6,500 ft. elev., Sept. 3, 1963 (Scullen and Bolinger); 2 ♂ ♂, 12 km. N. Rincon de Romos, Ags., July 28, 1951 (P. D. Hurd); 2 ♂ ♂, same locality and date (H. E. Evans); ♀, 41 mi. S. Saltillo, 6,200 ft. elev., Sept. 7, 1962 (U. of Kansas Mex. Exped.); 2 ♂ ♂. 5 mi. W. San Juan del Rio, Qro., 6,500 ft. elev., Sept. 2, 1963 (Scullen and Bolinger); 4 ♂ ♂, 10 mi. NE. of San Luis Potosi, 6,200 ft. elev., Aug. 22, 1954 (R. R. Dreisbach); ♂, 18 mi. W. San Luis Potosí, 6,000 ft. elev., Aug. 22, 1954 (J. G. Chillcott); 6 ♀ ♀, 21 ♂ ♂, 17 mi. NE. San Luis Potosí, 6,500 ft. elev., Oct. 3, 1957, at *Baccharis glutinosa* (H. A. Scullen); ♀, 18 mi. SW. San Luis Potosi, 7,300 ft. elev., Oct. 2, 1957 (H. A. Scullen); 5 ♀ ♀, 19 mi. SW. San Luis Potosi, 7,200 ft. elev., Sept. 4, 1963 (Scullen and Bolinger); 5 ♂ ♂, 4 mi. SW. San Luis Potosi, 6,500 ft. elev., Sept. 4, 1963 (Scullen and Bolinger; ♂ 40 mi. S. San Luis Potosi, 5,700 ft. elev., Sept. 5, 1963 (Scullen and Bolinger); 146 ♀ ♀, 411 ♂ ♂, 17 mi. NE. San Luis Potosi, 6,200 ft. elev., Sept. 6, 1963 (Scullen and Bolinger). NEW MEXICO: ♂, 7.5 mi. S. Three Rivers, Otero County, Sept. 9, 1961, at *Gutierrezia microcephala* (P. D. Hurd).

SPECIMENS OTHER THAN PARATYPES.—MEXICO: ♀, 10 mi. N. San Luis Potosi, 6,200 ft. elev., Aug. 22, 1954 (R. R. Dreisbach). TEXAS: ♂, Kent, Culberson County, 3,900 ft.–4,200 ft. elev., Sept. 17–18, 1912 (R. and H.); ♂, (location ?) 533 25 (V. E. Romney) (U.S.N.M.)

DISTRIBUTION.—South central plateau area of Mexico between 6,200 and 7,300 ft. elev. One record from New Mexico and one from Texas.

PREY RECORD.—None.

PLANT RECORD.—*Baccharis glutinosa* (Mexico). *Gutierrezia microcephala* (New Mexico). *Sphaeralcea* sp. (Mexico).

25. Eucerceris pacifica Scullen

FIGURES 31, 83 a, b, c

Eucerceris pacifica Scullen, 1948, pp. 156, 176–178, figs. 11A, B, C, 14.

Figure 31. Baja California. *E. pacifica* Scullen

TYPE.—The holotype male of *E. pacifica* Scullen from San Pedro, Baja California, Mexico, Oct. 7, 1941 (Ross and Bohart) is at the California Academy of Sciences. The female has not been identified.

DISTRIBUTION.—Known only from the type locality, and 10 mi. NW. of La Paz, Baja California.

PREY RECORD.—None.

PLANT RECORD.—Compositae (Baja California).

26. Eucerceris pimarum Rohwer

FIGURES 32, 84 a, b, c, d, e, f

Eucerceris pimarum Rohwer, 1908, pp. 326–327.—Scullen, 1965, pp. 135–136.
Eucerceris bitruncata Scullen, 1939, pp. 19, 35–36, figs. 24, 65, 98, 145; 1948, pp. 158, 171; 1951, p. 1011.
Eucerceris triciliata Scullen, 1948, pp. 156, 172–175, figs. 9A, 9B, 9C, 15; 1951, p. 1012.—Krombein, 1960a, pp. 77–79; 1960b, p. 300.—Evans, 1966, p. 146.

TYPES.—The holotype female of *E. pimarum* Rohwer is at the U.S. National Museum, no. 14110. It was taken at Phoenix, Ariz.

The holotype female of *E. bitruncata* Scullen is at the U.S. National Museum, no. 50835 and was taken at San Antonio, Tex., June 9, 1917 (J. C. Crawford). The holotype male of *E. triciliata* Scullen, is at the California Academy of Sciences and was taken 20 mi. N. of El Paso, Tex., June 19, 1942 (H. A. Scullen).

DISTRIBUTION.—Abundant throughout southern Arizona, southern New Mexico and western Texas, with scattered records in north central Mexico, south central Texas, southern California, southern Nevada, and southern Utah.

PREY RECORD.—*Minyomerus languidus* Horn (Krombein in Arizona, 1960a, b).

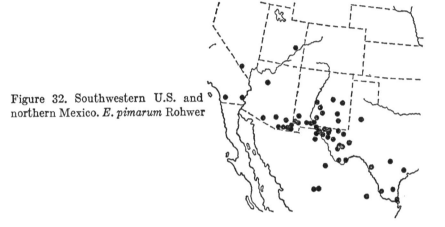

Figure 32. Southwestern U.S. and northern Mexico. *E. pimarum* Rohwer

PLANT RECORD.—*Acacia* sp. (Arizona, Texas). *Baccharis glutinosa* (Arizona, Texas). *Baileya pleniradiata* (New Mexico).*Bahia absinthifolia* (Arizona). *Chrysothamnus* sp. (California). *Cuscuta indecora* (Texas). *C. umbellata* (Arizona). *Eriogonum annum* (Texas). *E. rotundifolium* (New Mexico). *E. thomasii* (Arizona). *E. trichopes* (New Mexico). *E. wrightii* (New Mexico). *Euphorbia* sp. (New Mexico). *Gaillardia amblyodon* (Texas). *G. pulchella* (New Mexico). *Gutierrezia microcephala* (New Mexico). *Haplopappus* (*Aplopappus*) *hartwegi* (Arizona). *Hymenothrix wislizeni* (New Mexico). *Koeberlinia spinosa* (New Mexico). *Lepidium* sp. (New Mexico). *L. alyssoides* (Mexico). *Parthenium incanum* (Arizona). *Pectis papposa* (California, New Mexico). *Thelespermia megapotamicum* (Arizona). *Tidestromia lanuginosa* (Texas). *Verbesina encelioides* (Arizona).

27a. *Eucerceris punctifrons cavagnaroi*, new subspecies

FIGURE 33

Female.—Length 15 mm. Black with yellow markings; punctation small and crowded; pubescence short.

Head about one ninth wider than the thorax; black except for most of the face, large elongate patches back of the eyes, two large oval spots on the vertex back of the ocelli, most of the mandibles and a small patch on the scape, all of which are yellow; clypeal border with a medial rounded black extension, flanked on each side by a low black denticle on the medial lobe of the clypeus; mandibles with one large blunt denticle; antennae normal in form, ferruginous below and darker above.

Thorax black except for a divided band on the pronotum, a small patch on the posterior lobe, a trace on the mesoscutum, an emarginate band on the scutellum, the entire enclosure, a reversed J-shaped patch on the propodeum, two large angular patches on the mesopleuron and two small spots on the metapleuron, three pairs of patches on the sternum and a small patch on the tegula, all of which

Figure 33. Mexico. *E. punctifrons punctifrons* (Cameron), *E. punctifrons cavagnaroi*, new subspecies

are yellow; tegulae normal in form; the enclosure closely pitted except along the basal margin, and with a light medial groove; mesosternal tubercles absent; legs largely yellow but with dark fuscous areas on all coxae, trochanters, femora and on the mid and hind tibiae; wings light fulvous; 2nd submarginal cell not petiolate.

Abdomen with yellow bands on terga 1 to 5 deeply emarginate and the lateral expansion of the bands of terga 3, 4, and 5 with dark patches; a yellow patch on terga 6 laterad of the pygidium; first four sterna largely yellow; fifth sternum with lateral yellow areas; pygidium with sides convex, apex rounded, and base converging.

This unique female is structurally like the female of *E. punctifrons* Cameron except for the more extended yellow parts. In the nominate subspecies the yellow of the scutellum is separated into two lateral large spots, the enclosure is black except for two elongate yellow spots near the base, and the bands on the terga are much reduced.

MALE.—Unknown.

TYPE.—The holotype ♀ of *E. punctifrons cavagnaroi* Scullen was taken on Volcan de San Salvador, El Salvador, June 24, 1963, (D. Cavagnaro and M. E. Irwin). The mandibles of the type are badly worn and the apices of the wings are frayed. Deposited at the California Academy of Sciences.

DISTRIBUTION.—Known only from the type locality.

PREY RECORD.—None.

PLANT RECORD.—None.

27b. *Eucerceris punctifrons punctifrons* (Cameron)

FIGURES 33, 85 a, b, c

Aphilanthops punctifrons Cameron, 1890, p. 106, fig. 2 ♀ (♂ by error)
Cerceris punctifrons Kohl, 1890, p. 369.—Dalla Torre, C. G., 1897, p. 470.
Eucerceris punctifrons Scullen, 1939, pp. 18, 22–23, figs. 54, 55, 56, 135; 1948, p. 157.

TYPE.—The holotype female of *E. punctifrons* (Cameron) is at the British Museum of Natural History, no. 21.1,195. It was taken at Temax, North Yucatan, Mexico, by Gaumer.

DISTRIBUTION.—Known only from the holotype.

PREY RECORD.—None.

PLANT RECORD.—None.

28. *Eucerceris rubripes* Cresson

FIGURES 34, 86 a, b, c, d, e, f, g

Eucerceris rubripes Cresson, 1879, p. xxiii; 1882, pp. v, vi, vii; 1887, p. 281.—
    Ashmead, 1890, p. 32.—Bridwell, 1899, p. 209.—Ashmead, 1899, p. 295.—
    Viereck and Cockerell, 1904, pp. 84, 85, 88.—Viereck, 1906, p. 233.—Smith,
    H. S., 1908, p. 372.—Mickel, 1917, p. 455.—Cresson, 1916, p. 101.—Scullen,
    1939, pp. 18, 25–28, figs. 15, 16, 38, 60, 78, 93, 111, 125, 140; 1948, pp. 156,
    158; 1951, p. 1012.
Eucerceris unicornis Patton, 1879, pp. 359–360.—Cresson, 1887, p. 281.—
    Bridwell, 1899, p. 209.—Ashmead, 1899, p. 295.—Viereck and Cockerell,
    1904, pp. 84, 85, 87.—Viereck, 1906, p. 233.
Cerceris unicornis Schletterer, 1887, p. 505.—Dalla Torre, C. G., 1897, p. 480.
Cerceris rubripes Dalla Torre, K. W. von, 1890, p. 201.—Dalla Torre, C. G.,
    1897, p. 473.
Aphilanthops marginipennis Cameron, 1890, p. 105, t. 7, fig. 1.
Eucerceris marginipennis Kohl, 1890, p. 368.[1]
Cerceris marginipennis Dalla Torre, C. G., 1897, p. 467.

The female of *E. rubripes* Cresson is very close to the female of *E. apicata* Banks from which it may be separated by the following characteristics: (1) The mesal clypeal area of *rubripes* shows a dis-

---

[1] This reference is from Dalla Torre, 1897, p. 467. However, the original Kohl reference was examined and no reference was found to the species *Eucerceris marginipennis*. If Kohl used the above species name it must be in another publication so far undiscovered.

tinct dorsal carina with a medial single minute denticle just below the prominent acute tubercle common to both species; (2) The mandibles of *rubripes* are unidentate and the mandibles of *apicata* are bidentate; (3) The pygidium of *rubripes* is definitely rounded apically while in *apicata* it is truncate with the apical margin somewhat emarginate. The distribution of the two species overlap in the central Rocky Mountain area.

TYPES.—The lectotype male of *E. rubripes* Cresson from Colorado (H. K. Morrison) is in the Philadelphia Academy of Natural Sciences, no. 1961. The lectotype of *E. unicornis* Patton is at the Philadelphia Academy of Natural Sciences. The type male of *Aphilanthops*

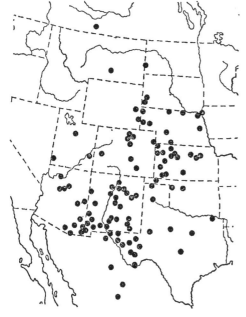

Figure 34. Western U.S. *E. rubripes* Cresson

marginipennis Cameron is at the British Museum of Natural History. The abdomen has been lost. It was taken at Atoyac in Vera Cruz, Mexico (Schumann).

DISTRIBUTION.—From southern Alberta S. through the Rocky Mts. into north central Mexico and E. to the Missouri River.

PREY RECORDS.—*Peritaxia* sp. (Cazier, 2 mi. NE. Portal, Ariz. July 28, 1961) (det. by Miss R. E. Warner).

PLANT RECORDS.—*Acacia angustissima* (Arizona). *Argemone* sp. (Nebraska). *Baccharis glutinosa* (Arizona, Mexico). *Baileya multiradiata* (Arizona). *B. pleniradiata* (New Mexico). *Compositae* (New Mexico). *Eriogonum* sp. (Arizona). *E. annum* (Texas). *E. thomasii* (Arizona). *Gutierrezia longifolia* (New Mexico). *Haploppapus hartwegi* (Arizona). *Helianthus* sp. (Nebraska, Mexico). *H. petiolaris* (North

Dakota). *Koeberlinia spinosa* (New Mexico). *Lepidium* sp. (Arizona, New Mexico). *L. montanum* (New Mexico). *Melilotus* sp. (New Mexico). *M. alba* (Nebraska). *Parthenium incanum* (Mariola) (Arizona, New Mexico). *Petalostemum* sp. (Nebraska). *Polygonum* sp. (Arizona, New Mexico). *Sartwellia mexicana* (New Mexico). *Solidago* sp. (Colorado, Nebraska). *Tamarix gallica* (New Mexico, Oklahoma, Texas). *Telesperma megapotamicum* (Arizona). *Xanthoxylum Clava-Herculis* (Texas).

29. Eucerceris ruficeps Scullen

FIGURES 35, 87 a, b, c, d, e, f

Eucerceris ruficeps Scullen, 1948, pp. 159, 175–176, ♀ figs. 10A, B, C, 14; 1951, p. 1012.—Linsley and MacSwain, 1954, pp. 11–14.—Krombein, 1958, p. 197; 1960a, pp. 78–79.

Figure 35. Southwestern U.S. *E. ruficeps* Scullen

At the time the female was described in 1948 the male of *E. ruficeps* Scullen was unknown. The male is described here. By error the holotype and paratypes were indicated as "males."

MALE.—Length 10 mm. Black with light yellow markings; punctation normal; pubescence short.

Head one fourth wider than the thorax, black except for the entire face and small narrow patches back of the eyes; small black patches appear above the antennal scrobes; three subequal denticles on the margin of the medial clypeal lobe; mandibles without denticles; antennae normal in form.

Thorax black except for the pronotum, a divided band on the scutellum, the metanotum, large patches on the propodeum, and most of the pleuron and ventral surfaces, all of which are light yellow;

tegulae low and smooth; enclosure with a light medial groove and heavily ridged subparallel to its base; mesosternal tubercle absent; legs all yellow except for variable patches of fulvous mesad on the femori and the slightly darkened tarsi; wings subhyaline with a clouded area over the anterior half; 2nd submarginal cell not petiolate.

Abdomen with broad yellow bands on terga 1 to 6, sterna 2 and 3, a small patch on sternum 1, an interrupted band on sternum 4, and lateral small patches on sternum 5, all of which are yellow; pygidium usual in form (fig. 87f).

TYPE.—The holotype female from Antioch, Calif., Aug. 7, 1938 (J. W. MacSwain) is at the California Academy of Sciences.

DISTRIBUTION.—A total of 20 females and 35 males is recorded from Antioch, Calif. One male was taken on Bethel Island near Antioch and one female was taken at Del Puerto Canyon in Stanislaus County, Calif. This species was found nesting at Antioch by Linsley and MacSwain (see Linsley and MacSwain, 1954). This nesting site has been taken over by industrial expansion and the known nesting site eliminated. A record from Johnnie, Nye County, Nev. is probably an error. Specimens are as follows:

CALIFORNIA: 3♀♀, 2♂♂, Antioch, Contra Costa County, May 18, 1936, June 8, 1933, June 1, 1939, Sept. 26, 1937 (G. E. and R. M. Bohart); ♀, ♂, same locality, May 21, 1936. Aug. 21, 1938 (E. C. Van Dyke); ♂, same locality, May 24, 1949 (E. G. Linsley); 2♀♀, 5♂♂, same locality, May 24, 1949, July 8, Aug. 9, 1947, Sept. 22, 1954 (P. D. Hurd); ♀, same locality, June 2, 1940 (B. Brookman); 6♂♂, same locality, June 15, 1952 (Cheng Liang, W. E. La Berge, L. D. Beamer, and A. E. Wolf); ♀, same locality, June 30, 1939 (R. M. Bohart); ♀, same locality, June 8, 1950 (J. E. Gillaspy); ♂, same locality, July 15, 1953 (S. Miyagawa); 7♂♂, same locality, July 16, 1953 (R. C. Bechtel, A. D. Telford, A. A. Grigarick); 2♂♂, same locality, July 31, 1959 (J. R. Powers, J. Powell); 2♀♀, ♂, same locality, Aug. 7, Sept. 4, 1938, Aug. 8, 1952 (J. W. MacSwain); ♀, ♂, same locality, Aug. 9, Sept. 10, 1947 (U. N. Lanham); 6♀♀, same locality, Aug. 9, Sept. 9, 1952 (E. G. Linsley and J. W. MacSwain); 2♂♂, same locality, Aug. 13, 1952 (W. F. Barr); 2♂♂, same locality, Aug. 15, 1954 (H. E. and M. A. Evans); ♀, same locality, Sept. 2, 1938 (G. E. Bohart); ♀, same locality, Sept. 9, 1954 (P. Torchio); ♀, Del Puerto Canyon, Stanislaus County, May 30, 1959 (F. D. Parker); ♂, Bethel Island, Contra Costa County, July 4, 1956. NEVADA.—♀, Johnnie, Nye County, July 1935.

BIOLOGY.—Linsley and MacSwain (1954, p. 11) found this species nesting in abandoned burrows of the Halictine bee, *Lasioglossum* (*Shpecodogastra*) *aberrans* (Crawford). The following prey records are from Linsley and MacSwain.

PREY RECORD.—*Dysticheus rotundicollis* Van Dyke (Antioch, Calif.). *Sitona californicus* Fahrens (Antioch, Calif).

PLANT RECORD.—*Croton californicus* (California).

30. Eucerceris similis Cresson

FIGURES 36, 88 a, b, c, d, e, f

Eucerceris similis Cresson, 1879, p. xxiv; 1882, pp. vi, vii; 1887, p. 281.—Ashmead, 1899, p. 295.—Cresson 1916, p. 101.—Scullen, 1939, pp. 18, 19, 30-32, figs. 20, 40, 62, 80, 95, 113, 127, 142; 1948, pp. 156, 159; 1951, p. 1012.
Cerceris similis Dalla Torre, K. W. von, 1890, p. 202.—Dalla Torre, C. G., 1897, p. 477.

TYPE.—The lectotype female of *E. similis* Cresson from Nevada (H. K. Morrison) is in the Philadelphia Academy of Natural History, no. 1965.1.

DISTRIBUTION.—*E. similis* Cresson is primarily a species of the mountains of California and southern Oregon with a few records from Nevada. Specimens are as follows:

Figure 36. Southwestern U.S. *E. similis* Cresson

130 females and 437 males have been recorded from California. These are mostly from central and northern localities of the state.

NEVADA: 3♂♂, Charcoal Ovens, State Park, White Pine County, July 5, 1960 (T. K. Haig); ♂, Galena, Washoe County, 6,300 ft. elev., July 26, 1959 (F. D. Parker); ♀, Glenbrook, Douglas County, Sept. 12, 1923 (Carl D. Duncan); 3♂♂, Nevada, "I-1880" (Morrison); ♀, 2♂♂, "Nevada." OREGON: 5♂♂, Colestin, Jackson County, July 30, 1918 (E. C. Van Dyke); ♀, Eagle Ridge, Klamath Lake, June 14, 1924 (C. L. Fox); 2♂♂, Green Spring Mt., Jackson County, 17 mi. E. of Ashland, Aug. 15, 1944 (H. A. Scullen); 3♀♀, 11♂♂, Lake-of-the-Woods, 4,950 ft. elev., July 18, 1930, Aug. 12, 1935 (H. A. Scullen); 2♀♀, 4♂♂, same locality, Aug. 12, 13, 1935 (George R. Ferguson); 11♂♂, Upper Klamath Lake, Klamath County, 4,200 ft. elev., Aug. 13, 1935, at *Daucus carota* (H. A. Scullen); 14♀♀, 21♂♂, W. side of Klamath Lake, Aug. 16, 1944, at *D. carota* (H. A. Scullen); 2♀♀, 3♂♂, Pelican Bay, Klamath Lake, July 22, 1930 (H. A. Scullen); 2♂♂, same locality, July 14, 1950 (C. Fitch); 4♀♀, 29♂♂, Prospect, Jackson Co., 2,578 ft. elev., Aug. 7, 9, 1944, July 27, 1945, at *Solidago* sp. (H. A. Scullen); ♀, 10 mi. N. of Prospect, Aug. 9, 1944 (H. A. Scullen).

PREY RECORD.—None.

PLANT RECORD.—*Achillea millefolium* (California). *Daucus carota* (Oregon). *Eriogonum nudum* (California). *Euphorbia serpyllifolia* (California). *Solidago* sp. (California, Oregon).

31. *Eucerceris sinuata* Scullen

FIGURES 37, 89 a, b, c, d, e, f, g, h

Eucerceris sinuata Scullen, 1939, pp. 18, 47, figs. 27, 70, 103, 150; 1948, p. 158; 1951, p. 1012; 1957, p. 155; 1964, pp. 205–208, figs. 1–4, map 1.
For a discussion of this species see the writer's 1964 paper.

TYPE.—The holotype female of *E. sinuata* Scullen taken at Devils River, Tex., May 5, 1907, on sumac (F. C. Bishopp), is in the U.S. National Museum, no. 50834.

Figure 37. Texas-Mexico Gulf Coast. *E. sinuata* Scullen

DISTRIBUTION.—South central Texas and northeastern Mexico. Specimens are as follows:

MEXICO: ♂, 4 mi. W. Linares, N. L., 1,300 ft. elev., Sept. 7, 1963 (Scullen and Bolinger); ♀, 11♂♂, Montemorelos, N. L. 1,700 ft. elev., Oct. 12, 1957 (H. A. Scullen); 11♂♂, same locality, Oct. 13, 1957 (H. A. Scullen); 4♀♀, 47♂♂, same locality, Sept. 8, 1963 (Scullen and Bolinger); 2♂♂, 23 mi. N. Sabinas, Coah., Aug. 10, 1959 (Menke and Stange). UNITED STATES: TEXAS: ♀, Devils River, May 5, 1907 (F. C. Bishopp); ♀, Leon Creek, Bexar County, Oct.17, 1952 (B. J. Adelson); ♀, same locality, Oct. 12, 1952 (M. Wasbauer).

PREY RECORD.—None.

PLANT RECORD.—*Baccharis glutinosa* (Mexico, Texas). Sumac (*Rhus* sp.) (Texas).

32. *Eucerceris sonorae* new species

FIGURES 38, 90 a, b, c, d, e, f, g

FEMALE.—Length 13 mm. Ferruginous to fuliginous and fuscous with light yellow markings; punctation average; pubescence average.

Head slightly wider than the thorax, ferruginous with a narrow darker patch bordering the eyes near the vertex and the apices of the mandibles and mandibular denticles which are dark fuscous (some specimens of females show dark stripes passing through the antennal scrobes fusing with a similar dark area embodying the ocelli); the following parts are yellow: large lateral eye patches, an elongate patch between the antennal scrobes, the entire clypeus except the border denticles and an evanescent patch back of the eyes; clypeal border with four subequal denticles, the medial pair of denticles connected by a concave lamella-like structure, a con-caved area between the lateral and the medial denticles, and a small cluster of bristles just above the medial pair of denticles; a low cone-shaped elevation appears on the surface of each lateral lobe of the clypeus; a small but distinct cone-shaped elevation with a truncate apex appears on the frons just above the epistomal suture; mandibles

Figure 38. Western Mexico. *E. sonorae*, new species

with one prominent acute denticle; antennae normal in form, fer-ruginous basally, becoming darker apically.

Thorax ferruginous infused with variable amounts of fuscous; yellow markings as follows: band on the pronotum extending onto the posterior lobe, evanescent lateral spots on the scutellum, the metanotum, angular patches on the pleuron, small patches on the tegulae, evanescent small spots on the enclosure, and large patches on the propodeum; tegulae low and smooth; enclosure smooth except for a medial groove and short ridges extending laterally from the meson; mesosternal tubercles small and acute but very distinct; legs ferruginous; wings subhyaline with the anterior area clouded; stigma and more basad border golden; second submarginal cell petiolate.

Abdomen ferruginous except for the depressed areas on terga 2, 3, and 4, and the following parts which are yellow: a broad band on terga 1 and 5, broad bands embodying the dark depressed areas on terga 2, 3, and 4, the sides of tergum 6, bands on sterna 3 and 4, and evanescent patches on sternum 5; pygidium with slightly convex sides converging to a small rounded apex.

MALE.—Length 13 mm. Ferruginous with limited areas darker and with light yellow markings; punctation average.

Head subequal in width to the thorax, ferruginous with a narrow darker patch bordering the eye near the vertex and the apices of the mandibles which are dark fuscous; face ferruginous except for broad lines through the antennal scrobes, an evanescent patch back of the eye, patch on base of mandible and spot on the scape, all of which are yellow; clypeal border with three denticles, the medial one the largest; mandibles without denticles; antennae normal in form, ferruginous basally with a trace of yellow on the scape and darker apically.

Thorax ferruginous becoming darker along the anterior and posterior margins of the mesoscutum and on the enclosure; light yellow markings as follows: band on prothorax extending onto the posterior lobe, angular patches on the pleuron, small patches on the tegulae, an evanescent band on the scutellum, the metanotum, and large patch on the propodeum; tegulae low and smooth; enclosure smooth except for a medial groove and short ridges extending out from the meson, mesosternal tubercle absent; legs ferruginous except for elongate patches on the fore and mid tibiae, and evanescent small patches on femora, all of which are yellow; wings subhyaline except for the usual clouded area along the anterior margin; 2nd submarginal cell not petiolate.

Abdomen ferruginous with wide yellow bands on terga 1 to 5, and sterna 3, 4, and 5, depressed area on tergum 2 small and ferruginous; depressed areas on terga 3, 4, and 5 fuscous; an evanescent yellow patch on sternum 1 and broken band of yellow on sternum 5, long rows of long loose bristles on sterna 3 and 4, a short row of short fused bristles on sternum 5; pygidium as illustrated (fig. 90g).

E. sonorae Scullen is very close to *E. canaliculata* (Say) in structure, size and color pattern. The females of the former species are easily distinguished by the prominent process on the frons just above the epistomal suture and the medial pair of denticles on the clypeal border which are relatively larger than on *E. canaliculata* (Say). The elevations on the lateral lobes of the clypeus of *sonorae* Scullen are relatively short compared to those on most females of *canaliculata* (Say). The males of *sonorae* Scullen and of *canaliculata* (Say) are indistinguishable except by association with the female.

TYPES.—The type ♀ and allotype ♂ of *E. sonorae* Scullen are from 32 mi. SE. of Guaymas, Son., Mexico, 125 ft. elev., Sept. 23, 1963

(H. A. Scullen and Duis Bolinger). They are deposited at the U.S. National Museum, no. 69230.

PARATYPES.—MEXICO: 15 ♀ ♀, 25 ♂ ♂, 32 mi. SE. of Guaymas, Son., Mexico, 125 ft. elev., Sept. 24, 1963 (H. A. Scullen and Dius Bolinger).

DISTRIBUTION.—Known only from the type location.

PREY RECORD.—None.

PLANT RECORD.—*Baccharis glutinosa.*

33. *Eucerceris stangei,* new species

FIGURES 39, 91 a, b, c, d, e, f, g, h, i, j

FEMALE.—Length 12 mm. Black with yellow markings; punctation small and crowded; pubescence general, of average length.

Head slightly wider than the thorax; black except for small eye patches, a small patch between the antennal scrobes and small spots back of the eyes, all of which are yellow; clypeal border with two prominent blunt extensions on the medial clypeal lobe between which there is a group of prominent bristles; clypeal surface without distinct elevations; a low cone-shaped elevation on the frons between the antennal scrobes; mandibles with a single prominent denticle curved ventrad, distad of which there is a concavity; antennae normal in form.

Thorax black, immaculate; pronotum with noticable elevations dorso-laterally; tegulae punctate over most of their surfaces; scutellum, metanotum and the entire propodeum, including the enclosure, closely and finely punctate; mesosternal tubercles absent; legs black except for an evanescent stripe on each tibia and a lighter area at the apical end of the 3rd femur; wings subhyaline but darker along the anterior half, 2nd submarginal cell petiolate.

Abdomen black except as follows: band on the apical half of tergum 3, two parallel bands on terga 4 and 5, tergum 6 exclusive of the pygidium, and broad bands on sterna 3, 4, and 5, all of which are yellow; pygidium with sides slightly convex and converging apically to a rounded apex, amber slightly infused with black.

MALE.—Length 12 mm. Black with yellow markings; punctation small and crowded; pubescence general, longer than average on many parts.

Head slightly wider than the thorax; black except for the entire face, basal three fourths of mandibles, anterior surface of the scape and a small spot back of the eye, all of which are yellow; clypeal surface depressed medially, border slightly extended on the medial lobe, margin without denticles; a low carina between the antennal scrobes; mandibles with an upper acute and a lower blunt denticle on the mesal surface, denticles connected by a curved ridge; antennae normal in form.

Thorax black, immaculate; tegulae punctate over their anterior half, otherwise smooth; scutellum, metanotum and the entire propodeum including the enclosure closely and finely punctate; mesosternal tubercles absent; legs black dorsally, yellow ventrally; wings subhyaline but darker along the anterior half; 2nd submarginal cell petiolate.

Abdomen black except for: band on the apical half of tergum 3 (some specimens have an evanescent narrow band on the basal half), two parallel bands on terga 4, 5, 6, and most of 7, all of which are yellow; pygidium slightly longer than broad, yellow infused with black; sterna 3, 4, 5, and 6 with variable amounts of yellow laterally;

[Figure 39. Mexican Gulf Coast. *E. stangei*, new species

ventral abdominal bristles forming distinct medially divided rows on sterna 3, 4, and 5, indistinct row on sternum 6; masses of prominent long setae covering much of sterna 3 and 4.

TYPES.—The type ♀ and allotype ♂ are from Mitla, Oaxaca, Mexico, 5,600 ft. elev., Aug. 20, 1963, on *Baccharis glutinosa* (H. A. Scullen and Duis Bolinger). Deposited at the U.S. National Museum, no. 69231.

PARATYPES.—MEXICO: 5♀♀, 4♂♂, Mitla, Oaxaca, Mexico, June 28, Aug. 20, 22, 1963, on *Baccharis glutinosa* (Scullen and Bolinger); 2♂♂, same locality, Sept. 2, 1957 (H. A. Scullen); ♀, 4. mi. NW. Tepanco de Lopez, Puebla, Mexico, July 2, 1953, on *Agave* (U. of Kansas Mex. Exped.); ♂, Cacaloapan, Puebla, Mexico, Aug. 20, 1963 (F. D. Parker and L. A. Stange).

DISTRIBUTION.—States of Oaxaca and Puebla, Mexico.

PREY RECORD.—None.

PLANT RECORD.—*Agave* (Puebla). *Baccharis glutinosa* (Oaxaca).

34a. *Eucerceris superba bicolor* Cresson

FIGURES 41, 92 a, b, c, d, e, f, g, h

Eucerceris bicolor Cresson, 1881, pp. xxxviii–xxxix; 1882, pp. v, vii; 1887, p. 281.—
Ashmead, 1899, p. 295.—Smith, H. S., 1908, pp. 371, 372.—Stevens, 1917,
p. 422.—Mickel, 1917, pp. 454, 455.—Carter, 1925, p. 133.—Cresson, 1916,
p. 99.
Eucerceris superba bicolor Scullen, 1939, pp. 18, 37–40; 1948, p. 158; 1951, p. 1012.
Cerceris dichroa Dalla Torre, K. W. von, 1890, p. 199.—Dalla Torre, C. G., 1897,
p. 457.

TYPE.—The holotype female of *E. bicolor* Cresson from Montana
(Morrison) is in the Philadelphia Academy of Natural Sciences, no.
1959.1.

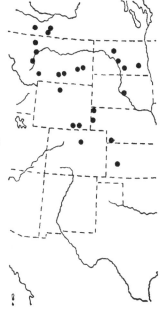

Figure 41. Western U.S. *E. superba*
superba Cresson

DISTRIBUTION.—Northern Rocky Mts. and western plains states
south to Colorado and Kansas. Specimens are as follows:

CANADA: ALBERTA: 3 ♀ ♀, Aug. 9, 1921, Aug. 7, 1923 (H. L. Seamans);
2 ♀ ♀, same locality, Aug. 5, 6, 1923 (H. E. Gray); ♀, same locality, Aug. 6, 1923
(Walter Carter); 2 ♀ ♀, same locality, Aug. 9, 1921, Aug. 17, 1939 (E. H. Strick-
land); 4 ♀ ♀, Manyberries, Aug. 11, 1939 (E. H. Strickland); 8 ♀ ♀, Medicine
Hat, Aug. 20, 1916, Aug. 1, 1917 (Prey: *Ophryastes* of the *sulcirostris-porosus*
complex, det. by W. J. Brown), (F. W. L. Sladen); ♀, same locality, July 18, 1929
(Owen Bryant); 23 ♀ ♀, same locality, Aug. 7, 1938, Aug. 13, 1939 (E. H.
Strickland); ♀, same locality (J. R. Mallock).
UNITED STATES: COLORADO: 3 ♀ ♀, Sunshine Canyon, Boulder County,
7,000 ft. (?) 1949. KANSAS: ♀, Finney County, Sept. 1894 (H. E. Menke). MON-

TANA: ♀, Bozeman, Aug. 24, 1911; ♀, Fallon, Prairie County, Aug. 8, 1962 (J. G. and B. L. Rosen); ♀, Forsyth, Aug. 11, 1903; ♀, Great Falls, July 18, 1936 (H. B. Hoeffler); ♀, Helena, Aug. 1906 (W. H. Mann); 2♀ ♀, Huntley, Aug. 16, 1916; ♀, Shelby, Sept. 6, 1924 (H. L. Seamans). NEBRASKA: ♀, Benkleman, July 1899 (L. Bruner); ♀, Glen, Sioux County, 1905; ♀, same locality, 4,000 ft. elev., Aug. 20, 1906 (H. S. Smith); ♀, Mitchell, July 26, 1916 (C. E. Mickel). NORTH DAKOTA: 2♀ ♀, Minot, Aug. 22, 1915, at *Kuhnistera oligophylla* (O. A. Stevens); 2♀ ♀, Steele, July 13, Aug. 4, 1919, at *K. oligophylla* (O. A. Stevens); 9♀ ♀, Valley City, Aug. 17, 23, 1917 (P. W. Fattig); ♀, Washburn, July 23, 1926 at *K. oligophylla* (O. A. Stevens). SOUTH DAKOTA: ♀, Pierre, 1949; 3♀ ♀, same locality (W. J. Fox); 5♀ ♀, same locality (no other data). WYOMING: ♀, Centennial, 1955 (D. Tyndall); ♀, Graybull, Aug. 16, 1927 (H. H. Knight); ♀, Laramie, 1955.

PREY RECORD.—*Ophryastes* of the *sulcirostris-porosus* complex (Medicine Hat, Alberta, Can.) (det. by W. J. Brown).

PLANT RECORD.—*Eriogonum annum* (North Dakota). *Kuhnistera oligophylla* (North Dakota). *K. purpurea* (North Dakota).

34b. *Eucerceris superba superba* Cresson

FIGURE 40

Eucerceris superbus Cresson, 1865, pp. 108-109.—Packard, 1866, p. 58.—Patton, 1879, pp. 356-357.—Cresson, 1882, pp. vi-vii; 1887. p. 281.—Ashmead, 1890, p. 32; 1899, p. 295.—Smith, H. S., 1908, p. 371.—Cresson, 1916, p. 101.—Stevens, 1917, p. 422.—Mickel, 1917, pp. 454, 456.—Carter, 1925, p. 133.—Scullen, 1939, pp. 18, 36–37, figs. 25, 25a, 42, 66, 83, 99, 115, 129, 146; 1948, pp. 156, 158; 1951, p. 1012.
Cerceris superba Schletterer, 1887, p. 503.—Dalla Torre, C. G., 1897, p. 478
Eucerceris fulviceps Cresson, 1879, p. xxiii; 1882, pp. v, vii; 1887, p. 281.—Ashmead, 1899, p. 295.—Cresson, 1916, p. 100.
Cerceris fulviceps Dalla Torre, K. W. von, p. 201.—Dalla Torre, C. G., 1897, p. 461.
Eucerceris fulviceps var. *rhodops* Viereck and Cockerell, 1904, pp. 84, 85, 88.—Cresson, Jr., 1928, p. 49.

Light vittae of face may or may not be convergent above antennae in the male of *E. superba* Cresson.

TYPE.—The lectotype male of *E. superba* Cresson from the Rocky Mts., Colorado Territory (Ridings) is in the Philadelphia Academy of Natural Sciences, no. 1967.1. The holotype female of *E. fulviceps* Cresson from New Mexico is in the Philadelphia Academy of Natural Sciences, no. 1960. The holotype female of *E. fulviceps rhodops* Viereck and Cockerell from Pecos, N. Mex., at flowers of *Eriogonum* sp. Aug. 19, 1903 (Wilmatte P. Cockerell) is in the Philadelphia Academy of Natural Sciences, no. 10397.

DISTRIBUTION.—This species is found in a limited way throughout the Rocky Mt. area and western plains from southern Canada to Arizona and New Mexico. Localities from which specimens have been taken:

ALBERTA: Barons, Lethbridge, Manyberries, Medicine Hat, Scandia, Steverville, Suffield. ARIZONA: Flagstaff, Mohave County, Prescott, South Rim of Grand Canyon, Williams. COLORADO: Boulder, Clear Creek, Cortez, Glenwood Springs, Great Sand Dunes, Gunnison, McElmo, Pingree Park, Red Wash, Sunshine Canyon (Boulder County). IDAHO: Downey, Hansen, Hazelton (Jerome County), King Hill, Lewiston, Oakley, Pocatello, Twin Falls, Teton. IOWA: Sioux City. KANSAS: Gray County. MINNESOTA: Browns Valley. MONTANA: Helena, West Yellowstone. NEBRASKA: Glen (Sioux County), Monroe Canyon (Sioux County). NEW MEXICO: Artesia, Glorieta, Jemez Springs, Koehler, Santa Fe, Sapello, Vaughn. NORTH DAKOTA: Beach, Dickinson, Minot, Steele, Washburn, Williston, Wirch. SOUTH DAKOTA: Fort Pierre, Philips, Pierre. UTAH: Clover, Cornish, Erda, Fillmore, Flowell, Howell, Morgan, Petersboro, Providence, St. John, Tooele, Tremonton, Utah Lake. WYOMING: Big Horn Basin, Graybull, Powell, Worland.

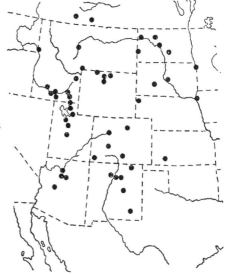

Figure 40. Western U.S. *E. superba bicolor* Cresson

PREY RECORD.—None.

PLANT RECORD.—*Cleome lutea* (Idaho). *Chrysothamnus* sp. (Utah). *Kuhnistera* (=*Petalostemon*) *oligophylla* (North Dakota). *Medicago sativa* (Alberta). *Melilotus alba* (Utah). *Solidago canadensis* (North Dakota).

35. *Eucerceris tricolor* Cockerell

FIGURES 42, 43, 93 a, b, c, d, e, f

Eucerceris vittatifrons tricolor Cockerell, 1897, p. 136.—Ashmead 1899, p. 295.
Eucerceris tricolor Viereck and Cockerell, 1904, pp. 84, 85, 87.—Scullen, 1939, pp. 17, 19, 53–54, figs. 32, 46, 73, 87, 106, 120, 134, 154; 1948, pp. 157, 158, 180; 1951, p. 1012.

TYPE.—The lectotype male of *E. tricolor* Cockerell taken at Las Cruces, N. Mex., 9–5 (Towns.) is in the Philadelphia Academy of Natural Sciences.

DISTRIBUTION.—This species is very abundant in southern Arizona, southern New Mexico, and western Texas. It has also been taken in the northern parts of Arizona and New Mexico. It is recorded from Mexico as given below.

CHIHUAHUA: ♀, 34 mi. S. of Chihuahua, 3,650 ft. elev., Oct. 25, 1957 (H. A. Scullen); 2♂♂, 10 mi. W. Jimenez, Sept. 11, 1950 (Ray F. Smith); ♀, Santa Clara Canyon, Parrita, 5 mi. W. of Chihuahua, July 5, 1954 (J. W. MacSwain). COAHUILA: 3♀♀, 4♂♂, 15 mi. N. Saltillo, 4,450 ft. elev., Sept. 9, 1963 (Scullen and Bolinger). HIDALGO: 4♂♂, "Rd. pass S. Cienega Pk.," Peloncillo Mts., 4,500 ft. elev., Aug. 27, 1937 (Rehn, Pate, Rehn). ZACATECAS: ♂, Fresnillo, 700 ft. elev., Aug. 15, 1947 (M. A. Cazier); ♂, Sain Alto, 7,000 ft. elev., Aug. 14, 1947 (M. A. Cazier).

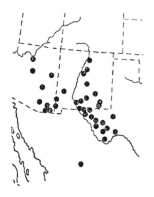

Figure 42. Southwestern U.S. and northern Mexico. *E. tricolor* Cockerell

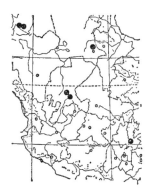

Figure 43. Central Mexico. *E. tricolor* Cockerell

PREY RECORDS.—None.

PLANT RECORDS.—*Acacia greggii* (Arizona). *Aplopappus* (*Haplopappus*) *hartwegi* (Arizona). *Baccharis glutinosa* (Arizona, Mexico, Texas). *Baileya multiradiata* (Arizona). *Chrysopsis hirsutissima* (New Mexico). *Condalia lycioides* (Arizona). Cotton (New Mexico). *Croton corymbulosus* (Texas). *C. texensis* (New Mexico). *Dithyera wislizeni* (New Mexico). *Eriogonum* sp. (Arizona). *E. abertianum neomexicanum* (Arizona). *E. annum* (Texas). *E. rotundifolium* (New Mexico). *Euphorbia capitellata* (Arizona). *E. pleniradiata* (Arizona). *Gutierrezia* sp. (Mexico). *G. microcephala* (Arizona). *Koeberlinia spinosa* (New Mexico). *Lepidium montanum* (Texas). *L. thurberi* (Arizona). *Melilotus alba* (Arizona). *Parthenium incanum* (Arizona). *Thelesperma megapotamicum* (Arizona).

36. *Eucerceris velutina* Scullen

FIGURES 44, 94 a, b, c, d, e, f, g

Eucerceris velutina Scullen, 1948, pp. 160–163, figs. 2A, B, C, D, E, F; 14

TYPES.—The type male and allotype female from San Bernardo, Mexico are at the California Academy of Sciences. The male was taken on Aug. 19, 1935 and the female on Aug. 16, 1935.

DISTRIBUTION.—In the mountainous area of Mexico from the state of Sonora to Oaxaca. Recorded only at or above 5,000 ft. elev. Specimens are as follows:

MEXICO: DURANGO: 2♀ ♀, 2♂ ♂, San Bernardo, Aug. 19, 1935. GUERRERO: ♀, 18 mi. S. Iguala, July 18, 1963 (Parker and Strange); ♀, Mexcala, June 29, 1951 (P. D. Hurd); ♀, Zumpango, July 22, 1963. GUANAJUATO: ♂, Gualajuato. JALISCO: ♀, Guadalajara (Crawford); 5♂ ♂, same locality, 5,000 ft. elev., July 14,

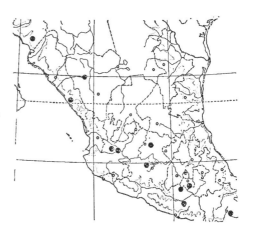

Figure 44. Central Mexico. *E. velutina* Scullen

1959 (H. E. Evans); ♀, 3 mi. SE. Plan de Barrancas, July 8, 1963 (Parker and Strange); 2♀ ♀, Teguila, July 19, 1954 (J. W. MacSwain). MICHOACAN: ♀, 11 mi. E. of Apatzigan, Aug. 20, 1954 (J. G. Linsley and J. W. MacSwain); ♂, 5 mi. E. Apatzigan, July 19, 1954 (J. G. Linsley and J. W. MacSwain); ♂, 6 mi. NW. Quiroga, July 11, 1963 (Parker and Stange). MORELOS 4♀ ♀, Tequesquitengo, July 15, 1961 (R. and K. Dreisbach). OAXACA: ♀, Mitla, 5,700 ft. elev. Sept. 2, 1957 (H. A. Scullen); 10♀ ♀, 12♂ ♂, same locality, June 27, 28, Aug. 20–22, 1963 (Scullen and Bolinger); 4♀ ♀, 10♂ ♂, 12 mi. SE. Oaxaca, 5,350 ft. elev., Aug. 21, 22, 1963 (Scullen and Bolinger); ♂, 23 mi. SE. Oaxaca, 5,600 ft. elev., June 28, 1963 (Scullen and Bolinger); ♀, 24 mi. SE. Oaxaca, Aug. 22, 1963 (Scullen and Bolinger). SINALOA: ♂, 8 mi. S. Elota, July 2, 1963 (F. D. Parker and L. A. Stange). SONORA: ♀, Minas Huevas, Aug. 7, 1952 (C. & P. Vaurie); ♂, 10 mi. E. Navajoa, Aug. 13, 1959 (W. L. Nutting, F. G. Werner); DURANGO: 2♀ ♀, 2♂ ♂, San Bernardo, Aug. 19, 1935.

PREY RECORD.—None.

PLANT RECORD.—*Baccharis glutinosa* (Oaxaca).

37. *Eucerceris violaceipennis* Scullen

FIGURES 45, 95 a, b, c

Eucerceris violaceipennis Scullen, 1939, pp. 18, 21–22, figs. 17, 58, 91, 138; 1948, p. 157.

TYPE.—The holotype female of *E. violaceipennis* Scullen from Cabima, Panama, May 21, 1911 (August Busck), is in the U.S. National Museum, no. 50836.

DISTRIBUTION.—This species is known only from the type locality in Panama. This unique female is of special interest for the two reasons mentioned by this writer in his 1939 paper. It is the most

Figure 45. Panama. *E. violaceipennis* Scullen

southern record for the genus and it is one of three species in which the 2 submarginal cell of the fore wing is not petiolate. Efforts have been made to locate the "Cabima" recorded as the type location. The late J. Zetek, for many years in charge of the Balboa Laboratory, was most helpful in this connection. Several localities carry the name "Cabima" so that it has been impossible to determine without question which is the true type locality. Several of the localities by that name are now under water. In view of the fact that it is a very large species measuring 23 mm. in length, it is strange no other collector has taken it.

PREY RECORD.—None.

PLANT RECORD.—None.

38. *Eucerceris vittatifrons* Cresson

FIGURES 46, 96 a, b, c, d, e, f

Eucerceris vittatifrons Cresson, 1879, p. xxiv; 1882, pp. vii, viii; 1887, p. 281.—
   Ashmead, 1899, p. 295.—Cresson, 1916, p. 101.—Scullen, 1939, pp. 17, 19,
   51-53, figs. 31, 49, 72, 86, 105, 119, 133, 153; 1948, pp. 157, 158, 180; 1951,
   p. 1013.
Cerceris vittatifrons Dalla Torre, K. W. Von, 1890, p. 202.—Dalla Torre, C. G.,
   1897, p. 481.

TYPE.—The lectotype male of *E. vittatifrons* Cresson from Nevada
(H. K. Morrison) is in the collection at the Philadelphia Academy
of Natural Sciences, no. 1969.1.

The female of *E. vittatifrons* Cresson is very close to the female

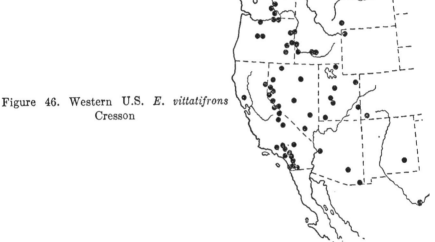

Figure 46. Western U.S. *E. vittatifrons* Cresson

of *E. arenaria* Scullen but they may be separated by the following
characters: The elevation on the medial clypeal surface is more
prominent and pointed on the latter than on the former, and the
basal end of the pygidium of the latter is not constricted as it is in
the former.

DISTRIBUTION.—All Pacific western states. Specimens from Mexico
are as follows:

MEXICO: AGUASCALIENTES: ♂, 12 km. N. Rincon de Romos, July 28, 1951
(H. E. Evans). SAN LUIS POTOSÍ: 8♂ ♂, 10 mi. NE. San Luis Potosi, 6,200 ft.
elev., Aug. 22, 1954 (Dreisbach); ♂ 84 mi. NE. San Luis Potosi, 4,000 ft. elev.,
Aug. 22, 1954 (Dreisbach); 3♂ ♂, 18 mi. W. San Luis Potosi, 6,000 ft. elev.,
Aug. 21, 1954 (J. G. Chilcott).

PREY RECORD.—None.

PLANT RECORD.—*Chrysothamnus nauseosus* (Nevada). *C. viscidiflorus typius* (California). *Eriogonum* sp. (Arizona, California, Oregon). *E. gracile* (California). *Melilotus alba* (Nevada, Oregon). *Solidago* sp. (Oregon). *Helianthus annuus* (Oregon). Yarrow (*Achillea*) (Nevada).

39. Eucerceris zimapanensis, new species

FIGURES 47, 97 a, b, c

MALE.—Length 14 mm. Black and ferruginous with yellow markings; punctation average; pubescence short.

Head subequal in width to the thorax; background color black except for a large dark ferruginous area on the gena and yellow which covers the face exclusive of the black vitta through the antennal scrobes; clypeal border with three denticles, the medial one the

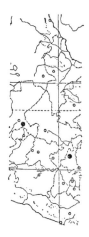

Figure 47. Central Mexico. *E. zimapanensis,* new species

largest; mandibles without denticles, ferruginous with darker apices; antennae normal in form, basal segments ferruginous, apical segments fuscous.

Thorax black except for the band on the pronotum extending onto the posterior lobe, an angular patch on the pleuron, two small lateral spots on the scutellum, the metanotum, large patches on the propodeum and a patch on the tegula, all of which are yellow, ferruginous areas include most of the scutellum and the borders of most of the yellow markings including the tegulae; tegulae normal in form; enclosure largely smooth but with a medial groove and short adjoining ridges and with a few scattered lateral punctures; mesosternal tubercles lacking; legs ferruginous with a slightly darker area on the hind tibiae, small yellow spots on the fore coxae and limited fuscous on all coxae; wings subhyaline with a darker area along the anterior portion, 2nd submarginal cell not petiolate.

Abdomen with background and depressed areas black to ferruginous with a broad emarginate band on terga 1 and 6, divided bands on terga 2, 3, 4, and 5, a divided small patch on sternum 2, an irregular small band on sternum 3 and a broken band on sternum 4, all of which are yellow; wide bands of long bristles on sterna 3 and 4, a short band of closely appressed short bristles on sternum 5; pygidium as illustrated (fig. 97c).

FEMALE.—Unknown.

The male of *E. zimapanensis* Scullen superficially looks very much like the male of *E. mellea* Scullen but the latter has only a single small divided cluster of bristles on sternum 5.

TYPE.—The ♂ type of *E. zimapanensis* Scullen was taken 9 mi. N. of Ojo Caliente, Zac., Mexico, May 12, 1962 (F. D. Parker). It is deposited at the California Academy of Sciences.

PARATYPES.—MEXICO: ♂, 9 mi. N. Ojo Caliente, Zac., May 19, 1962 (F. D. Parker); 2♂♂, Zimapan, Hdgo., June 11–14, 1951 at *Eysenhardtia polystachya* (Ort.).

DISTRIBUTION.—Known only from the type locality and from Zimapan, Hdgo.

PREY RECORD.—None.

PLANT RECORD.—*Eysenhardtia polystachya*. (Ort.).

40. *Eucerceris zonata* (Say)

FIGURES 48, 98 a, b, c, d, e, f, g, h

Philanthus zonata Say, 1823, p. 80. ♂; 1828, pl. xlix.—LeConte, 1883, vol. I, pp. 111–112, 167.
Eucerceris zonata Cresson, 1865, pp. 105–107.—Packard, 1866, p. 58.—Cresson, 1872, p. 227.—Patton, 1880, p. 398.—Cresson, 1882, pp. vi–vii; 1887, p. 281.—Robertson, 1892, pp. 104, 107; 1894, pp. 453–455, 458, 460; 1896, p. 73.—Ashmead, 1899, p. 295.—Smith, John B., 1900, p. 519.—Viereck and Cockerell, 1904, pp. 85, 88.—Smith, John B., 1910, p. 678.—Mickel, 1917, pp. 454, 456.—Washburn, 1919, p. 219, pl. 2, fig. 8.—Britton, 1920, p. 341.—Robertson, 1928, pp. 13, 24, 55, 68, 70–72, 90, 92, 107, 115, 120, 122, 123, 153–155, 195, 198.—Scullen, 1939, pp. 18, 40–43, figs. 26, 44, 67, 84, 100, 116, 130, 147; 1948, pp. 156, 158; 1951, p. 1013.
Eucerceris laticeps Cresson, 1865, pp. 107–108, ♂, ♀.—Packard, 1866, pp. 58–59.—Patton, 1879, p. 357.—Cresson, 1887, p. 281.—Ashmead, 1899, p. 295.—Smith, John B., 1900, p. 519; 1910, p. 678.—Leonard, 1928, p. 1017.—Cresson, 1916, p. 100.
Cerceris zonata Schletterer, 1887, p. 506.—Dalla Torre, K. W. von, 1890, p. 200.—Dalla Torre, C. G., 1897, p. 481.
Cerceris laticeps Schletterer, 1887, p. 495.—Dalla Torre, C. G., 1897, p. 466.
Eucerceris zonata var. *laticeps* Cresson, 1882, p. vii.—Smith, Harry S., 1908, pp. 371, 372.

TYPES.—The neotype female of *E. zonata* (Say) from Illinois is in the collection of the Philadelphia Academy of Natural Sciences. The holotype female of *E. laticeps* Cresson from Massachusetts is at the Philadelphia Academy of Natural Sciences, no. 1962.1.

DISTRIBUTION.—This is the only species recorded east of the Mississippi River. It has been taken from the New England states west to the Dakotas, eastern Wyoming, and Colorado. It is recorded from Arkansas and northern Texas but not south of the Ohio River east of the Mississippi River.

PREY RECORD.—None.

PLANT RECORD.—*Aster multiflorus* (Nebraska). *Eupatorium serotium* (Illinois). *Helianthus petiolaris* (North Dakota). *Petalostemum violaceus* (Nebraska). *Solidago* sp. (Indiana). *S. serotina* (North Dakota). *Vernonia fasciculata* (Nebraska).

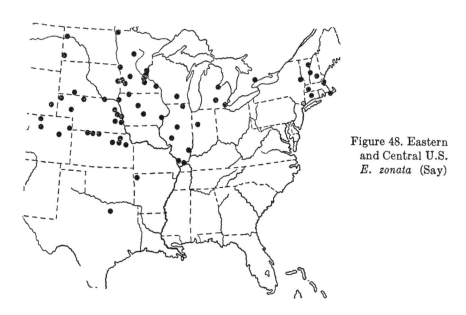

Figure 48. Eastern and Central U.S. *E. zonata* (Say)

Species of *Eucerceris*

Valid names appear without parentheses; synonyms with; names of species changed in status or changed in name from the original by this or previous publications are in brackets. Page numbers of principal references are in italics.

1. *angulata* Rohwer, 1912, 6, 7, *9–11*
2. *apicata* Banks, 1915, 5, 7, *11–12*, 27, 56
3. *arenaria* Scullen, 1948, 3, 7, *13–15*
   (*arizonensis* Scullen, 1939), 34, 35
4. *atrata* Scullen, new species, 4, 8, *15–17*
5. *baccharidis* Scullen, new species, 6, 8, *17–19*
6. (*baja* Scullen, 1948), 5, *19*, 20

28. *rubripes* Cresson, 1879, 2, 4, 7, *55–57*
29. *ruficeps* Scullen, 1948, 2, 5, 9, *57–58*
30. *similis* Cresson, 1879, 5, 8, *59–60*
     (*simulatrix* Viereck and Cockerell, 1904), 33
31. *sinuata* Scullen, 1939, 4, 8, 43, *60*
32. *sonorae* Scullen, new species, 5, 7, *60–63*
     (*sonorensis* Cameron, 1891) [*Cerceris*], 45
33. *stangei* Scullen, new species, 6, 7, 43, *63–64*
     (*striareata* Viereck and Cockerell, 1904), 31
34a. *superba bicolor* Cresson, 1881 [*C. bicolor* Cresson, 1881], 8, *65–66*
34b. *superba superba* Cresson, 1865, 4, 8, 65, *66–67*
     (*tricilliata* Scullen, 1948), 2, 52
35. *tricolor* Cockerell, 1897 [*C. vittatifrons* var. *tricolor* Cockerell
     1897], 5, 9, *67–68*
     (*unicornis* Schletterer, 1879), 55, 56
36. *velutina* Scullen, 1948, 3, 6, *69*
37. *violaceipennis* Scullen, 1939, 6, *70*
38. *vittatifrons* Cresson, 1879, 6, 7, *71–72*
     [*vittatifrons* var. *tricolor* Cockerell, 1897], 67
39. *zimapanensis* Scullen, new species, 4, *72–73*
40. *zonata* (Say), 1823, 4, 8, *73–74*
     (*zonata* var. *laticeps* Cresson, 1882), 73

Beetle Prey of *Eucerceris*

CURCULIONIDAE

Dyslobus lecontei Casey. Wasp: *E. flavocincta* Cresson_ _ _ _ _ _ _ _p. 2, 32
Dyslobus segnis (LeConte). Wasp: *E. flavocincta* Cresson_ _ _ _ _ _ _p. 32
Dysticheus rotundicollis Van Dyke. Wasp: *E. ruficeps* Scullen_ _p. 2, 58
Minyomerus languidus Horn. Wasp: *E. pimarum* Rohwer_ _ _ _p. 2, 53
Ophryastes sulcirostris (Say). Wasp: *E. superba bicolor* Cresson_ _p. 66
Sitona californicus Fahrens. Wasp: *E. ruficeps* Scullen_ _ _ _ _ _ _ _ _p. 58
Peritazia sp. Wasp: *E. rubripes* Cresson_ _ _ _ _ _ _ _ _ _ _ _ _ _ _ _ _ _p. 2, 56

Literature Cited

ASHMEAD, W. H.
    1890.   On the Hymenoptera of Colorado. Colorado Biol. Assoc. Bull., no. 1,
            pp. 1–47.
    1899.   Classification of the Entomophilous wasps, or the superfamily
            Sphecoidea. Paper no. 5. Family XX, Philanthidae. Canad. Ent.,
            vol. 31, pp. 291–300.
BANKS, NATHAN
    1915.   New fossorial Hymenoptera. Canad. Ent., vol. 47, pp. 400–406.
BOHART, R. M. and POWELL, J. A.
    1956.   Observations on the nesting habits of *Eucerceris flavocincta* Cresson
            (Hymenoptera: Sphecidae). Pan-Pacific Ent., vol. 32, pp. 143–144.

BRADLEY, J. C.
1921.   Some features of the Hymenopterous fauna of South America. Actes de la Socété Scientifique de Chile, vol. 33, pp. 51–74.

BRIDWELL, J. C.
1899.   A list of Kansas Hymenoptera. Trans. Kansas Acad. Sci., vol. 16, pp. 203–211.

BRITTON, WILTON E.
1920.   Check-list of the insects of Connecticut. Conn. State Geological and Natural History Survey, Bull. 31, p. 341.

CAMERON, PETER
1890–1. Biologia Centralis-Americana. Hymenoptera, vol. 2, pp. 103–134, pls. 7, 8.

CARTER, WALTER
1925.   Records of Alberta Sphecoidea with descriptions of new species of Crabronidae. Canad. Ent., vol. 57, pp. 131–136.

COCKERELL, T. D. A.
1897.   New Hymenoptera from New Mexico. The Entomologist, vol. 30, pp. 135–138.

CRESSON, E. T., SR.
1865.   Monograph of the Philanthidae of North America. Proc. Ent. Soc. Philadelphia, vol. 5, pp. 85–132.
1872.   Hymenoptera Texana. Trans, Amer. Ent. Soc., vol. 4, pp. 153–292.
1879.   [Notes on Hymenoptera.] Trans. Amer. Ent. Soc., vol. 7, pp. xxiii–xxiv.
1881.   [Notes on Hymenoptera.] Trans. Amer. Ent. Soc., vol. 9, pp. xxxviii–xxxix.
1882.   [Notes on Hymenoptera.] Trans. Amer. Ent. Soc., vol. 10, pp. v–viii.
1887.   Synopsis of the families and genera of the Hymenoptera of America, north of Mexico, together with a catalogue of the described species and bibliography. Trans. Amer. Ent. Soc., suppl. vol., pp. 1–351.
1916.   The Cresson types of Hymenoptera. Amer. Ent. Soc. Mem., no. 1, pp. 1–141.

CRESSON, E. T., JR.
1928.   The types of Hymenoptera in the Academy of Natural Sciences of Philadelphia other than those of Ezra T. Cresson. Mem. Amer. Ent. Soc., no. 5, pp. 1–90.

DALLA TORRE, C. G.
1897.   Catalogus Hymenopterorum hucusque descriptorum systematicus et synonymicus. Cerceris, vol. 8, pp. 449–481.

DALLA TORRE, K. W. VON
1890.   Hymenopterologische Notizen. Wien. Ent. Zeit., vol. 9, pp. 199–204.

EVANS, HOWARD E.
1966.   The behavior patterns of solitary wasps. Ann. Rev. Ent., vol. 11, pp. 123–154.

GAHAN, A. B. and ROHWER, S. A.
1917–8. Lectotypes of the species of Hymenoptera (except Apoidea) described by L'Abbé L. Provancher. Canad. Ent., vol. 49, pp. 298–308, 331–336, 391–400, 427–433; vol. 50, pp. 28–33, 101–106, 133–137, 166–171, 196–201.

GROSSBECK, JOHN A.
1912.   List of insects collected by the "Albatross" expedition in Lower
California in 1911, with description of a new species of wasp.
[*Eucerceris angulata* Rohwer.] Bull. Amer. Mus. Nat. Hist., vol.
31, pp. 323–326.
KOHL, FRANZ FRIEDRICH
1890.   Die Hymenopterengruppe der Sphecinen I. Monographie der
Naturlichen Gattung *Sphex* Linne. Ann. des Wiener Museums
des Naturghistorisches, vol. 5, pp. 77–194, 317–462. (See footnote
no. 1, p. 55)
1896.   Die Gattungen der Sphegiden. Annalen des K. K. Naturhisto-
rischen Hofmuseums. Wien, vol. 11, pp. 233–516.
KROMBEIN, KARL V.
1958.   Hymenoptera of America north of Mexico. U.S. Dept. of Agric.
Monogr., no. 2, suppl. 1, p. 197.
1960a.   Biological notes on several southwestern ground-nesting wasps
(Hymenoptera: Sphecidae). Bull. Brooklyn Ent. Soc., vol. 55,
pp. 75–79.
1960b.   Life history and behavioral studies of solitary wood- and ground-
nesting wasps and bees in southeastern Arizona. Year Book
Amer. Philo. Soc., 1960, pp. 299–300.
LE CONTE, JOHN L. (ed.)
1883.   The complete writings of Thomas Say on the entomology of North
America, vol. 1, pp. 1–54.
LEONARD, M. D.
1928.   A list of the insects of New York. New York (Cornell) Agric. Exp.
Stat., Mem. 101, pp. 1–1121.
LINSLEY, E. G., and MACSWAIN, J. W.
1954.   Observations on the habits and prey of *Eucerceris ruficeps* Scullen
(Hymenoptera: Sphecidae). Pan-Pacific Ent., vol. 30, pp. 11–14.
MICKEL, CHARLES E.
1916.   New species of Hymenoptera of the superfamily Sphecoidea. Trans.
Amer. Ent. Soc., vol. 42, pp. 399–434.
1917.   A synopsis of the Sphecoidea of Nebraska (Hymenoptera). Nebraska
Univ. Stud., vol. 17, pp. 342–456.
PACKARD, A. S.
1866–7. Revision of the fossorial Hymenoptera of North America, I: Crabronidae and Nyssonidae. Proc. Ent. Soc. Philadelphia, vol. 6,
pp. 39–115, 353–444.
PATE, V. S. L.
1937.   The generic names of the sphecoid wasps and their type species.
(Hymenoptera: Aculeata). Amer. Ent. Soc. Mem., no. 9, pp. 1–103.
PATTON, W. H.
1879.   List of a collection of aculeate Hymenoptera made by S. W.
Williston in northwestern Kansas. Bull. U.S. Geol. Geogr. Surv.
Terr., vol. 5, no. 3, pp. 349–370.
1880.   Notes on the Philanthinae. Proc. Boston Soc. Nat. Hist., vol. 20,
pp. 397–405.
PROVANCHER, L'ABBE L.
1889.   Additions et corrections au vol. II, faune entomologique du Canada:
Tritant des Hyménoptères. Quebec, pp. 418–419.

ROBERTSON, CHARLES
1892.   Flowers and insects: Labiatae. Trans. Acad. Sci., St. Louis, vol. 6, pp. 101–131.
1894.   Flowers and insects: Rosaceae and Compositae. Trans. St. Louis Acad. Sci., vol. 6, pp. 435–480.
1896.   Flowers and insects, XV. Bot. Gaz., vol. 21, pp. 72–81.
1928.   Flowers and insects. Privately printed, Carlinville, Ill. pp. 1–221.

ROHWER, S. A.
1908.   New Philanthid wasps. Canad. Ent., vol. 40, pp. 322–327.
1912.   In: Grossbeck, List of insects of Lower California. Bull. Amer. Mus. Nat. Hist., vol. 31, pp. 323–326.

SAY, THOMAS
1923.   A description of some new species of Hymenopterous insects. Western Quart. Reporter, vol. 2, no. 1, pp. 71–82.
1828.   Amer. Ent., vol. III.

SCHLETTERER, AUGUST
1887.   Die Hymenopteren-Gattung Cerceris Latr. mit vorzugsweiser Berücksichtigung der paläarktischen Arten. Zoologische Jahrbucher, Zeitschr. System, vol. 2, pp. 349–510, pl. XV.

SCHULZ, W. A.
1906.   Spolia Hymenopterologica. Paderborn, pp. 77–269.

SCULLEN, HERMAN A.
1939.   A review of the genus Eucerceris (Hymenoptera: Sphecidae). Oregon State Mono., Stud. Ent., no. 1, pp. 1–80, figs. 1–158 e.
1948.   New species in the genus Eucerceris with notes on recorded species and a revised key to the genus. Pan-Pacific Ent., vol. 24, pp. 155–180, figs. 1–15.
1951.   Tribe Cercerini. In: Muesebeck, Krombein and Townes, Hymenoptera of America north of Mexico. U.S. Dept. Agric., Agric. Mono. 2, pp. 1004–1013.
1957.   Cercerini collection notes, I. Pan-Pacific Ent., vol. 33, pp. 155–156.
1959.   Eucerceris simulatrix Viereck and Cockerell misspelled on type label. Ent. News., vol. 70, p. 108.
1961.   Synonymical notes on the genus Cerceris, III. (Hymenoptera: Sphecidae). Pan-Pacific Ent., vol. 37, pp. 45–49.
1964.   The male of Eucerceris sinuata Scullen (Hymenoptera: Sphecidae). Ent. News, vol. 75, pp. 205–208, map 1, figs. 1–4.
1965a. Revised synonymy in the genus Eucerceris with a description of the true female of E. elegans Cresson (Hymenoptera: Sphecidae). Ent. News, vol. 76, pp. 131–136.
1965b. Review of the genus Cerceris in America North of Mexico. Proc. U.S. Nat. Mus., vol. 116, no. 3506, pp. 333–548.

SMITH, HARRY S.
1908.   The Sphegoidea of Nebraska. Nebraska Univ. Stud., vol. 8, pp. 323–411.

SMITH, JOHN B.
1900.   Insects of New Jersey. New Jersey State Board Agric., 27th Ann. Rep., 1899, suppl.
1910.   A report of the insects of New Jersey. New Jersey State Mus. Ann. Rep., 1909, pp. 14–888.

SNOW, F. H.
1881.   Preliminary list of Hymenoptera of Kansas. Trans. Kansas Acad. Sci., vol. 7, pp. 97–101.

STEVENS, O. A.
1917. Preliminary list of the North Dakota wasps exclusive of Eumenidae. Ent. News, vol. 28, pp. 419–423.
VIERECK, HENRY L.
1902. Hymenoptera from southern California and New Mexico, with descriptions of new species. Proc. Acad. Nat. Sci. Philadelphia, vol. 54, pp. 728–743.
1906. Notes and descriptions of Hymenoptera from the Western United States. Trans. Amer. Ent. Soc., vol. 32, pp. 173–247.
VIERECK, H. L. and COCKERELL, T. D. A.
1904. The Philanthidae of New Mexico, I. New York Ent. Soc. Journ. vol. 12, pp. 84–88.
WASHBURN, F. L.
1918. The Hymenoptera of Minnesota. State Entomol. Minnesota, 17th Rep. (April 20, 1919), vol. 17, pp. 145–237, figs. 1–125.

Abbreviations

Apx_3	Reniform apex of hind femur
ATP	Anterior tentorial pits
CxC_1, CxC_2, CxC_3	Coxal cavities
Enc	Enclosure of propodium
Eps_1, Eps_2	Episternum
Es	Epistomal suture
Mes_1, Mes_2	Mesosternum (2 plates)
Met_1, Met_2	Metasternum (3 plates)
Mpct	Mesopectus
MsT	Mesosternal tubercle
N_1, N_3	Pronotum, Metanotum
Pl_3	Metaplura
PLP	Posterior lobe of pronotum
Prop	Propodeum
Pyg	Pygidium
S_1, S_2, S_3	Sterna
Sc_2, Sc_3	Submarginal cells
Scl_2	Scutellum
Sct	Scutum
Tg	Tegula
T_1, T_2, T_3	Terga

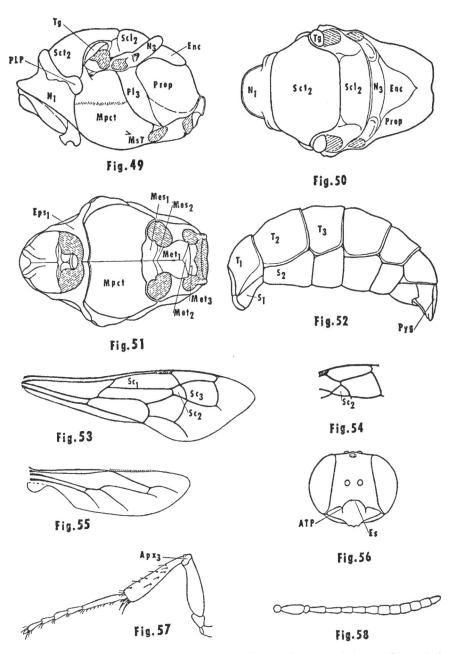

FIGURES 49-58.—49, Lateral aspect of thorax; 50, dorsal aspect of thorax; 51, ventral aspect of thorax; 52, lateral aspect of abdomen of male; 53, forewing; 54, apical cells of forewing with petiolate second submarginal cell; 55, posterior wing; 56, face of male; 57, leg showing reniform distal end of hind femora; 58, normal form of antennae of male.

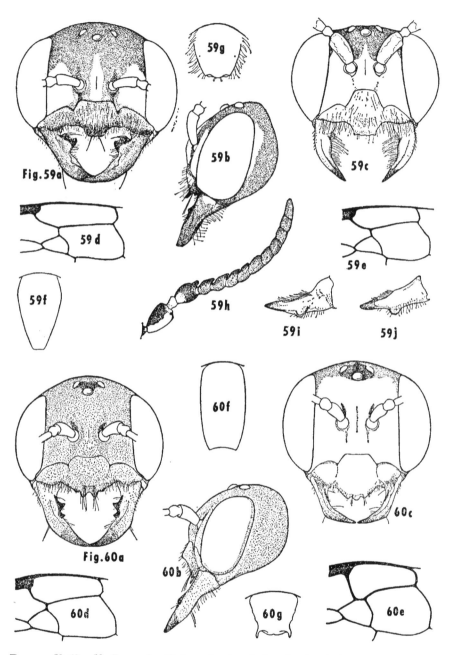

FIGURES 59–60.—59, *E. angulata* Rohwer (a=female face, b=female head profile, c=male face, d=female wing, e=male wing, f=female pygidium, g=male pygidium, h=male antennae, posterior view, i=male mandible, dorsal, j=male mandible, ventral); 60, *E. apicata* Banks (a=female face, b=female head profile, c=male face, d=female wing, e=male wing, f=female pygidium, g=male pygidium).

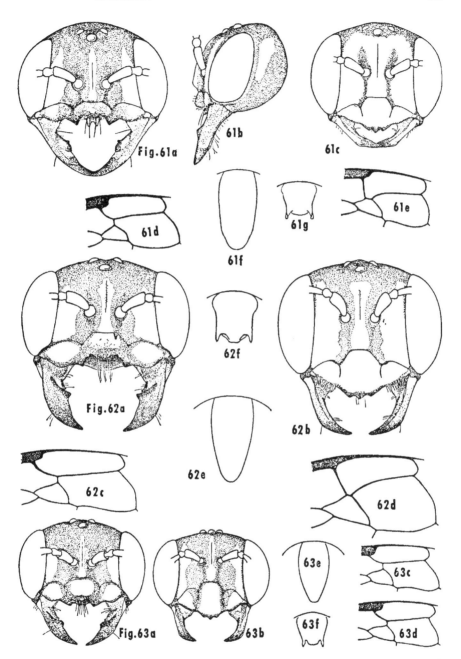

FIGURES 61–63.—61, *E. arenaria* Scullen (a=female face, b=female profile, c=male face, d=female wing, e=male wing, f=female pygidium, g=male pygidium); 62, *E. atrata* Scullen (a=female face, b=male face, c=female wing, d=male wing, e=female pygidium, f=male pygidium); 63, *E. baccharidis* Scullen (a=female face, b=male face, c=female wing, d=male wing, e=female pygidium, f=male pygidium).

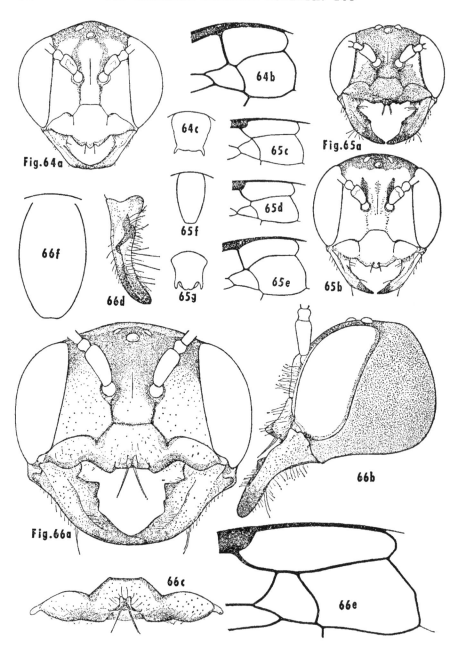

FIGURES 64–66.—64, *E. baja* Scullen (a=male face, b=male wing, c=male pygidium); 65, *E. barri* Scullen (a=female face, b=male face, c=female wing and d=female wing showing variation, e=male wing, f=female pygidium, g=male pygidium); 66, *E. brunnea* Scullen (a=female face, b=female profile, c=female clypeus, d=female mandible, mesal view, e=female wing, f=female pygidium).

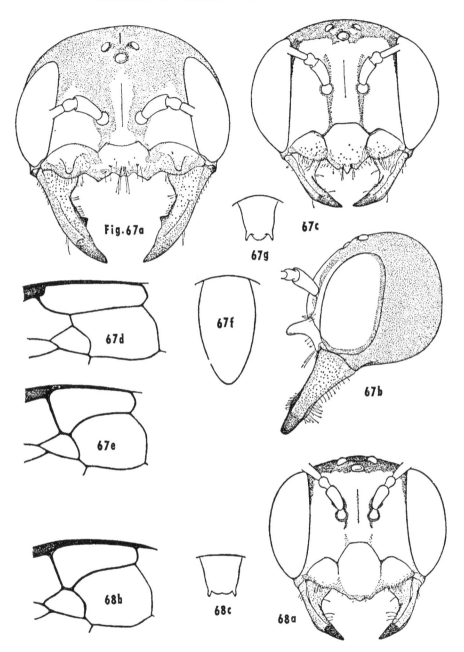

FIGURES 67–68.—67, *E. canaliculata canaliculata* (Say) (a=female face, b=female profile, c=male face, d=female wing, e=male wing, f=female pygidium, g=male pygidium); 68, *E. elegans elegans* Cresson (a=male face, b=male wing, c=male pygidium).

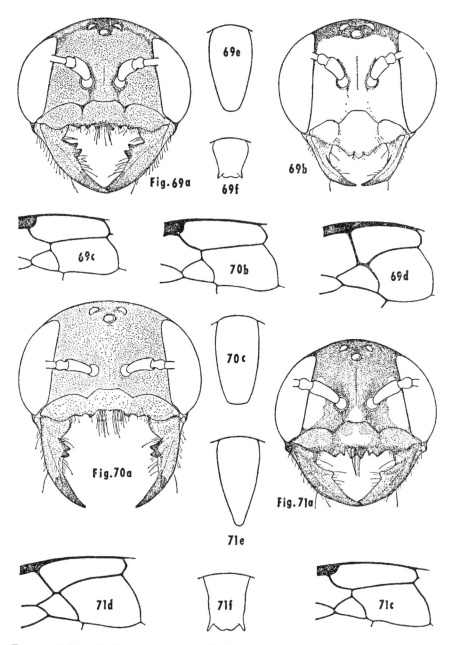

FIGURES 69–71.—69, *E. elegans monoensis* Scullen (a=female face, b=male face, c=female wing, d=male wing, e=female pygidium, f=male pygidium); 70, *E. ferruginosa* Scullen (a=female face, b=female wing, c=female pygidium); 71, *E. flavocincta* Cresson (a=female face [b=male face, page 87] c=female wing, d=male wing, e=female pygidium, f=male pygidium).

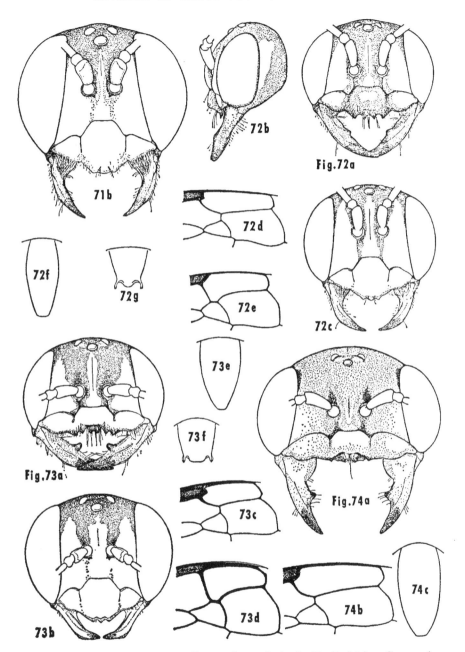

FIGURES 71–74.—71, *E. flavocincta* Cresson (b=male face); 72, *E. fulvipes* Cresson (a= female face, b=female profile, c=male face, d=female wing, e=male wing, f=female pygidium, g=male pygidium); 73, *E. insignis* Provancher (a=female face, b=male face, c=female wing, d=male wing, e=female pygidium, f=male pygidium); 74, *E. lapazensis* Scullen (a=female face, b=female wing, c=female pygidium).

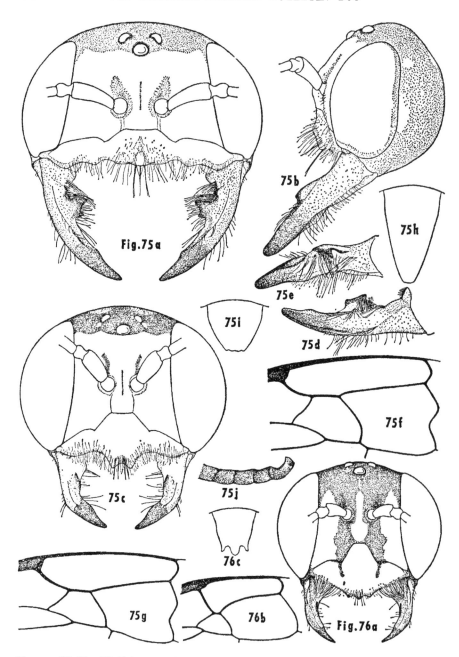

FIGURES 75–76.—75, *E. lacunosa* Scullen (a = female face, b = female profile, c = male face, d = female mandible, lateral view, e = female mandible, mesal view, f = female wing, g = male wing, h = female pygidium, i = male pygidium, j = male antennae); 76, *E. melanosa* Scullen (a = male face, b = male wing, c = male pygidium).

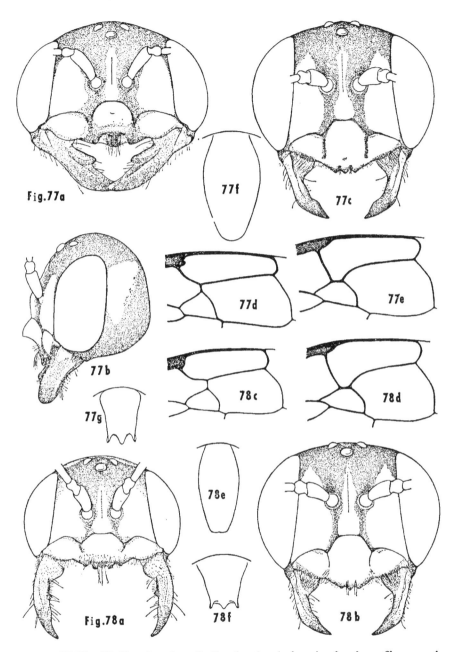

FIGURES 77–78.—77, *E. melanovittata* Scullen (a=female face, b=female profile, c=male face, d=female wing, e=male wing, f=female pygidium, g=male pygidium); 78, *E. mellea* Scullen (a=female face, b=male face, c=female wing, d=male wing, e=female pygidium, f=male pygidium).

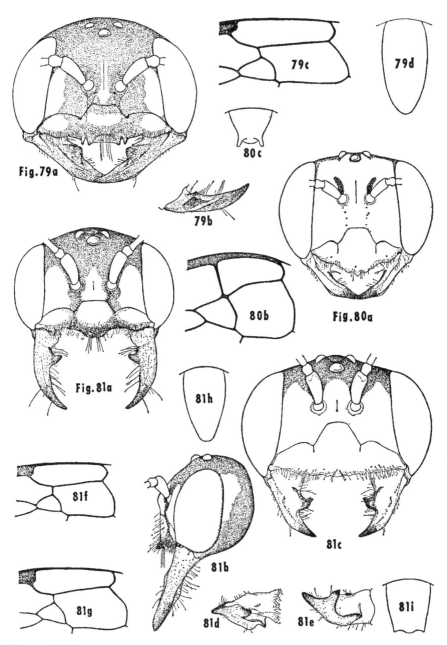

FIGURES 79–81.—79, *E. menkei* Scullen (a=female face, b=female mandible, c=female wing, d=female pygidium); 80, *E. mojavensis* Scullen (a=male face, b=male wing, c=male pygidium); 81, *E. montana* Cresson (a=female face, b=female profile, c= male face, d=male mandible, mesad, e=male mandible, ventral, f=female wing, g= male wing, h=female pygidium, i=male pygidium).

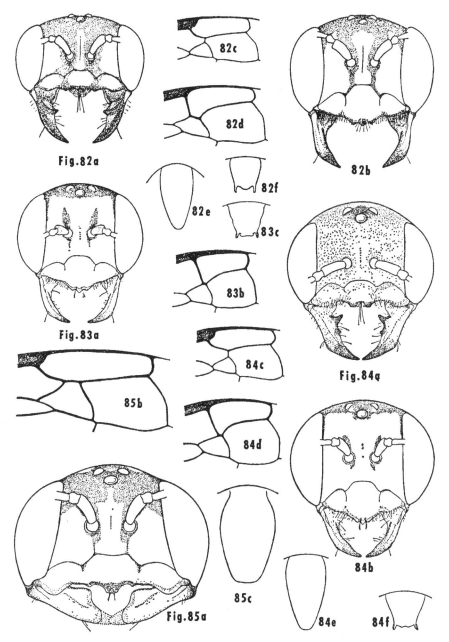

FIGURES 82–85.—82, *E. morula* Scullen (a=female face, b=male face, c=female wing, d=male wing, e=female pygidium, f=male pygidium); 83, *E. pacifica* Scullen (a=male face, b=male wing, c=male pygidium); 84, *E. pimarum* Rohwer (a=female face, b=male face, c=female wing, d=male wing, e=female pygidium, f=male pygidium); 85, *E. punctifrons* (Cameron) (a=female face, b=female wing, c=female pygidium).

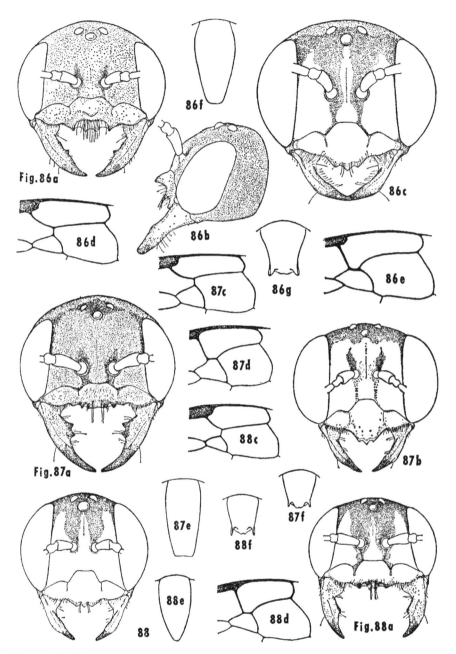

FIGURES 86–88.—86, *E. rubripes* Cresson (a=female face, b=female profile, c=male face, d=female wing, e=male wing, f=female pygidium, g=male pygidium); 87, *E. ruficeps* Scullen (a=female face, b=male face, c=female wing, d=male wing, e=female pygidium, f=male pygidium); 88, *E. similis* Cresson (a=female face, b=male face, c=female wing, d=male wing, e=female pygidium, f=male pygidium).

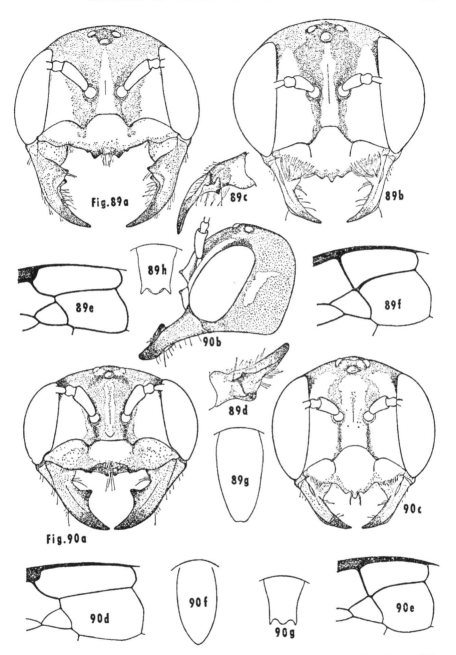

FIGURES 89–90.—89, *E. sinuata* Scullen (a=female face, b=male face, c=female mandible, dorsal, d=female mandible, mesal, e=female wing, f=male wing, g=female pygidium, h=male pygidium); 90, *E. sonorae* Scullen (a=female face, b=female profile, c=male face, d=female wing, e=male wing, f=female pygidium, g=male pygidium).

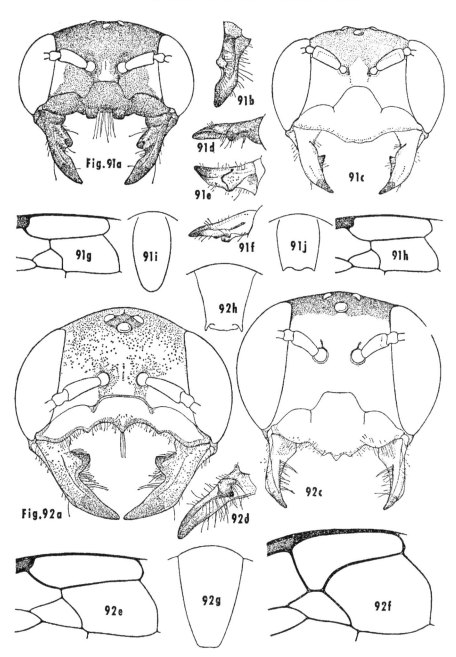

FIGURES 91-92.—91, *E. stangei* Scullen (a=female face, b=female mandible, ventral view, c=male face, d=male mandible, ventral view, e=male mandible, ventro-mesal view, f=male mandible, dorso-mesal view, g=female wing, h=male wing, i=female pygidium, j=male pygidium); 92, *E. superba* Cresson (a=female face, [b=female profile, page 95], c=male face, d=female mandible, e=female wing, f=male wing, g=female pygidium, h=male pygidium).

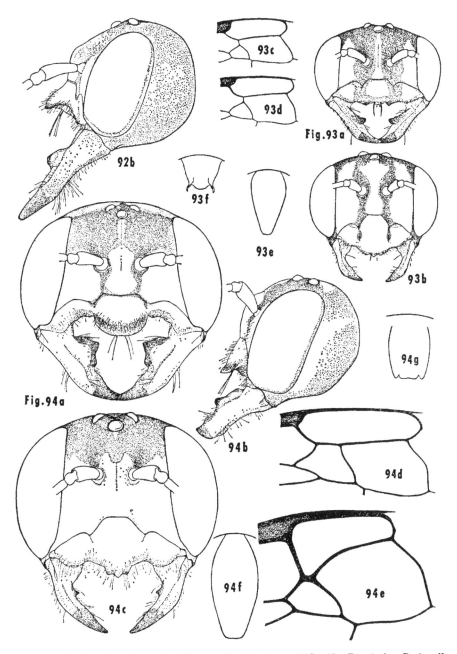

FIGURES 92–94.—92, *E. superba* Cresson (b=female profile); 93, *E. tricolor* Cockerell (a=female face, b=male face, c=female wing, d=male wing, e=female pygidium, f=male pygidium); 94, *E. velutina* Scullen (a=female face, b=female profile, c=male face, d=female wing, e=male wing, f=female pygidium, g=male pygidium).

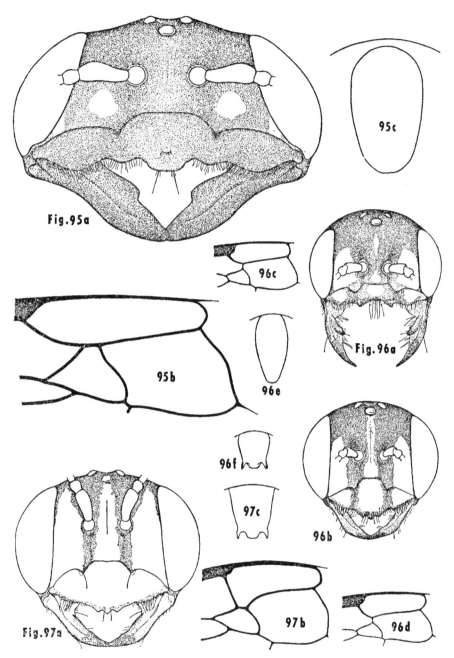

FIGURES 95–97.—95, *E. violaceipennis* Scullen (a=female face, b=female wing, c=female pygidium); 96, *E. vittatifrons* Cresson (a=female face, b=male face, c=female wing, d=male wing, e=female pygidium, f=male pygidium); 97, *E. zimapanensis* Scullen (a=male face, b=male wing, c=male pygidium).

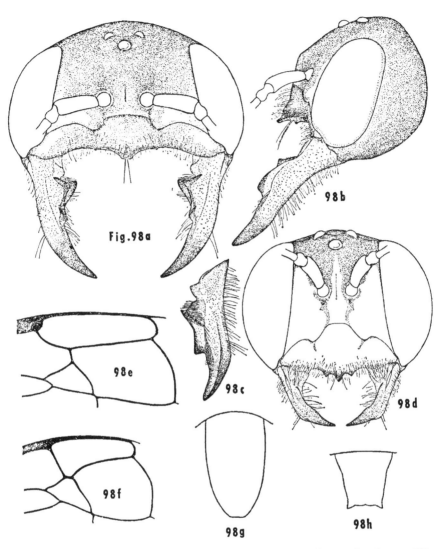

FIGURE 98.—*E. zonata* (Say) (a=female face, b=female profile, c=female mandible, d=male face, e=female wing, f=male wing, g=female pygidium, h=male pygidium).

CPSIA information can be obtained
at www.ICGtesting.com
Printed in the USA
BVHW04*1210180918
527831BV00013B/891/P